THE PERSHING II FIRING BATTERY

HEADQUARTERS
DEPARTMENT OF THE ARMY

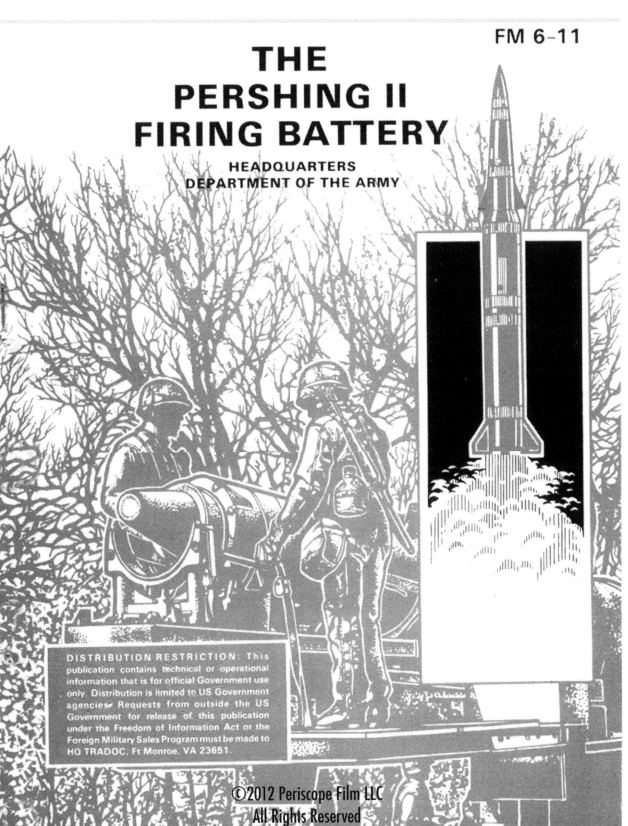

DISTRIBUTION RESTRICTION: This publication contains technical or operational information that is for official Government use only. Distribution is limited to US Government agencies. Requests from outside the US Government for release of this publication under the Freedom of Information Act or the Foreign Military Sales Program must be made to HQ TRADOC, Ft Monroe, VA 23651.

FM 6-11

Field Manual
Number 6-11

HEADQUARTERS
DEPARTMENT OF THE ARMY
Washington, DC, 13 March 1985

THE PERSHING II
FIRING BATTERY

TABLE OF CONTENTS

Pershing II is a terminally guided theater fire support missile system that can accurately deliver tactical nuclear warheads. It has the longest range and the greatest payload of any US Army weapon system. The system is ground mobile, air transportable, and can be employed worldwide.

This publication describes doctrine, techniques, considerations, and planning factors to maintain combat readiness and sustain combat operations within the Pershing II firing battery. Guidance is provided on the duties of key personnel and the training of sections, platoons, and the battery as a whole.

All users of this publication are encouraged to recommend changes and improvements. Comments should be keyed to the specific page, paragraph, and line of the text in which the change is recommended. Rationale should be provided in support of comments to ensure complete understanding and evaluation. Comments should be prepared, using DA Form 2028 (Recommended Changes, to Publications and Blank Forms), and forwarded directly to:

Commandant
US Army Field Artillery School
ATTN: ATSF-WGP
Fort Sill, Oklahoma 73503-5600

FM 6-11 is designed for use in conjunction with user's manuals for the appropriate equipment and with other technical manuals as referenced throughout this publication.

When used in this publication, "he," "him," "his," and "men" represent both the masculine and feminine genders unless otherwise stated.

CHAPTER 1
SYSTEM DESCRIPTION

Pershing II (PII) is a ground-mobile, surface-to-surface, nuclear weapon system. It is a solid-propellant missile with ground support equipment (GSE) mounted on *wheeled* vehicles. The missile may be launched quickly and is an effective weapon against a broad spectrum of targets.

MISSILE CHARACTERISTICS

The Pershing II missile, with the normal configuration of first- and second-stage propulsion sections and the reentry vehicle, weighs more than *16,000 pounds*, is about 10.6 meters long, and has a range of 1,000 miles (1,800 km). A kit is available with which the erector-launcher can be converted to adapt to a single-stage missile, if required. The result is a missile with reduced weight, length, and range.

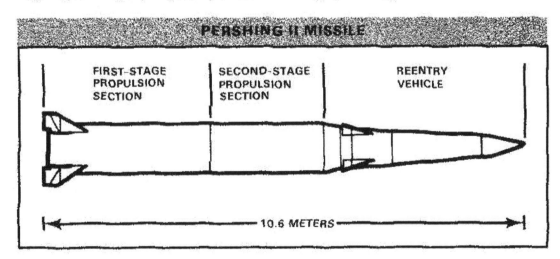

PERSHING II MISSILE

FIRST-STAGE PROPULSION SECTION

SECOND-STAGE PROPULSION SECTION

REENTRY VEHICLE

10.6 METERS

FIRST-STAGE
PROPULSION SECTION

The first stage provides the initial thrust required to propel the missile toward a ballistic trajectory. It consists of a rocket motor with attached forward and aft skirts. The forward skirt allows the second stage to be mated to the first stage. In a single-stage configuration, the reentry vehicle is mated directly to the forward skirt. The forward skirt contains three thrust termination ports. These ports stop the forward thrust at the time determined by the guidance and control section of the reentry vehicle when the missile is fired in the single-stage configuration. The first stage contains a motor ignition safe/arm mechanism designed to prevent the accidental launching of the missile. The aft skirt contains four air vanes (two fixed and two movable) and a swivel nozzle, which provide pitch and yaw control. The motor cases of both the first and second stages are made of Kevlar, a strong, lightweight material.

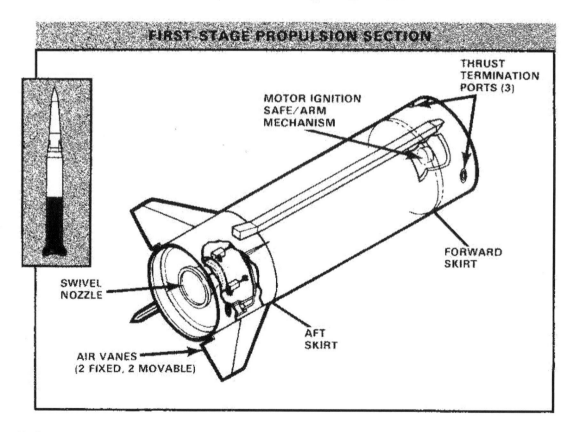

FIRST-STAGE PROPULSION SECTION

SECOND-STAGE
PROPULSION SECTION

The second stage produces thrust for a variable amount of time and accelerates the missile to the required programed velocity for achieving target range. Like the first stage, the second stage consists of a solid-propellant rocket motor with attached forward and aft skirts. The forward skirt contains three thrust termination ports, which stop the forward thrust at the time determined by the guidance and control section of the reentry vehicle. The forward skirt also allows the stage to be mated to the reentry vehicle. The aft skirt houses a swivel nozzle with two hydraulic actuators for nozzle control. It provides a means for mating the first and second stages. The aft skirt also contains a linear shaped charge to cause first-stage separation.

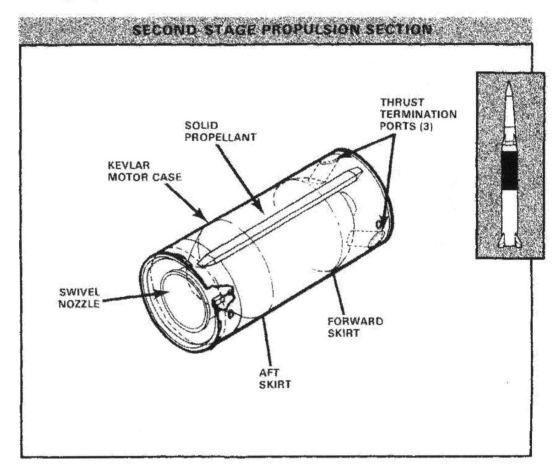

SECOND-STAGE PROPULSION SECTION

REENTRY VEHICLE

The reentry vehicle (RV) consists of three sections: a guidance and control/adapter (G&C/A) section, a warhead section, and a radar section.

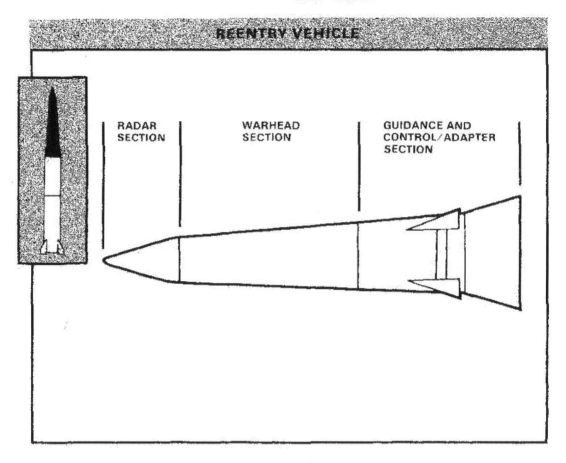

The *guidance and control/adapter* contains:

■ The terminal guidance system.

■ Thrusters to provide attitude control while in less dense atmosphere.

■ Air vanes for roll control during second-stage thrust and full control once the reentry vehicle returns to denser atmosphere.

■ The adapter portion, which is designed so the reentry vehicle can be mated to either the first- or second-stage propulsion section. The adapter remains attached to the propulsion section when the reentry vehicle separates from it.

GUIDANCE AND CONTROL/ADAPTER SECTION

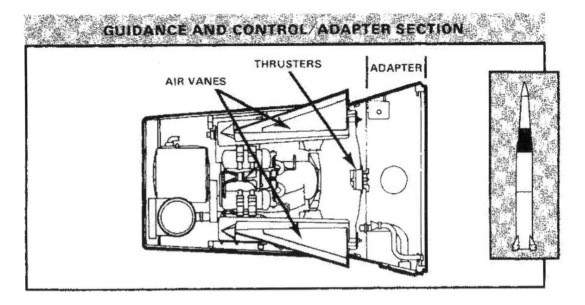

The *warhead section* is a conical aluminum structure coated with an ablative material. It houses:

■ A nuclear warhead that can provide an airburst, a surface burst, or an air/surface burst.

■ A permissive action link (PAL) unlocking device.

■ The rate gyro, which sends trajectory information to the G&C/A section.

WARHEAD SECTION

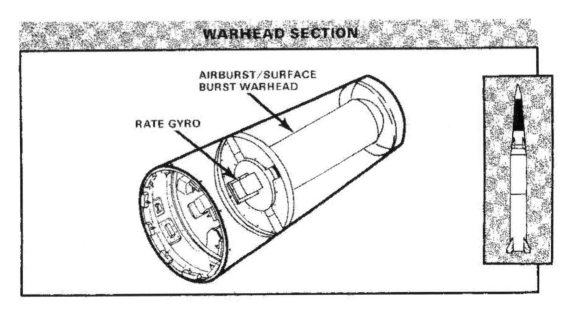

The *radar section*, consisting of a radome assembly, a radar unit, and a stabilized antenna unit, is located in the nose of the reentry vehicle. It transmits radio frequency (RF) energy to the target area, receives altitude and video return, and sends this altitude information to the on-board computer in the G&C/A section.

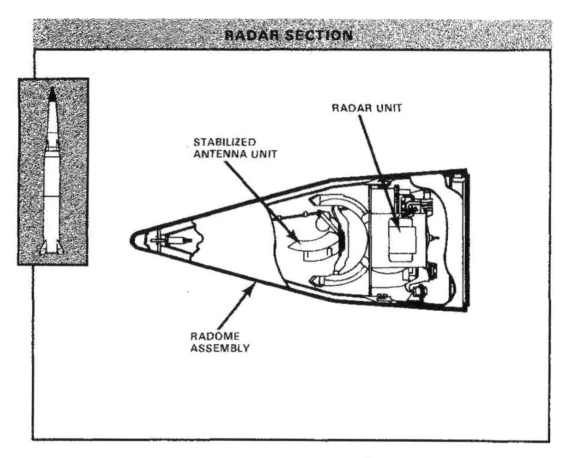

GROUND SUPPORT EQUIPMENT

To be effectively employed, the Pershing missile requires certain system-peculiar ground support equipment. This equipment is designed to provide speed, flexibility, reliability, and improved survivability.

ERECTOR-LAUNCHER

Each missile is fired from an erector-launcher (EL). The EL consists of a transporter frame assembly that supports an erection system, a leveling system, a radar section/warhead section assembly and transport pallet, a reentry vehicle cooling system, a ground integrated electronic unit (GIEU), a hydraulic control panel, and two 28-volt DC power supplies. Through the GIEU, the EL operator can test and monitor EL functions and, when required, can control the countdown. The GIEU also provides launch site communications between the erector-launcher and the platoon control central.

ERECTOR-LAUNCHER

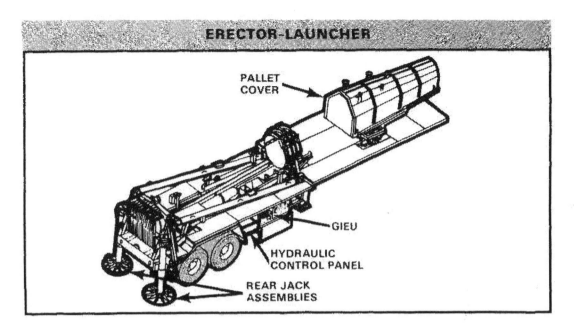

PALLET COVER

GIEU

HYDRAULIC CONTROL PANEL

REAR JACK ASSEMBLIES

PRIME MOVER

The prime mover for the erector-launcher is a 10-ton tractor with a self-recovery winch. Mounted on this tractor are the 30-kw generator, a power distribution box, a crane with a telescopic extension, and hydraulic stabilizer jacks.

TRACTOR WITH CRANE AND 30-KW GENERATOR

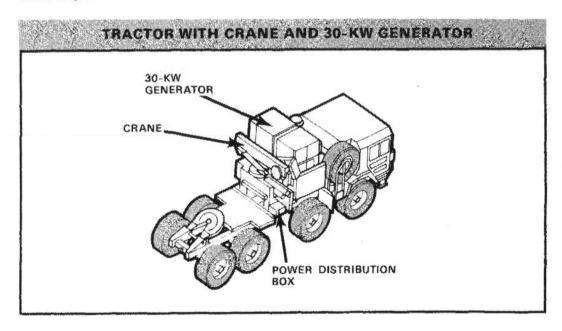

30-KW GENERATOR

CRANE

POWER DISTRIBUTION BOX

PLATOON CONTROL CENTRAL

Each firing platoon has a platoon control central (PCC). It is mounted on a 5-ton, long-wheelbase, cargo truck. The power source for the PCC is the trailer-mounted 30-kw generator. The PCC serves as the platoon command post, communications center, and launch control facility. Three remote launch control units (RLCU) are located in the PCC. They provide PAL functions, safe/arm functions, and firing control for up to three missiles on launchers. Next to the RLCUs is an interface logic assembly (ILA) console. The ILA contains three keyboards for manual data entries, three alphanumeric displays for detailed countdown status information, and three line printers that provide hard copies of countdown data.

The PCC also contains a launch window status display panel and a missile status display panel. The launch window status display panel computes time limits within which the platoon may launch its missiles. The missile status display panel gives the officer in charge (OIC) current missile countdown status information as it is developed by the ILA console.

PLATOON CONTROL CENTRAL WITH TRAILER-MOUNTED 30-KW GENERATOR

ANTENNA MAST BASES

PCC SHELTER

CHEMICAL/BIOLOGICAL PROTECTIVE ENTRANCE

30-KW GENERATOR

SIGNAL ENTRY PANEL

POWER ENTRY PANEL

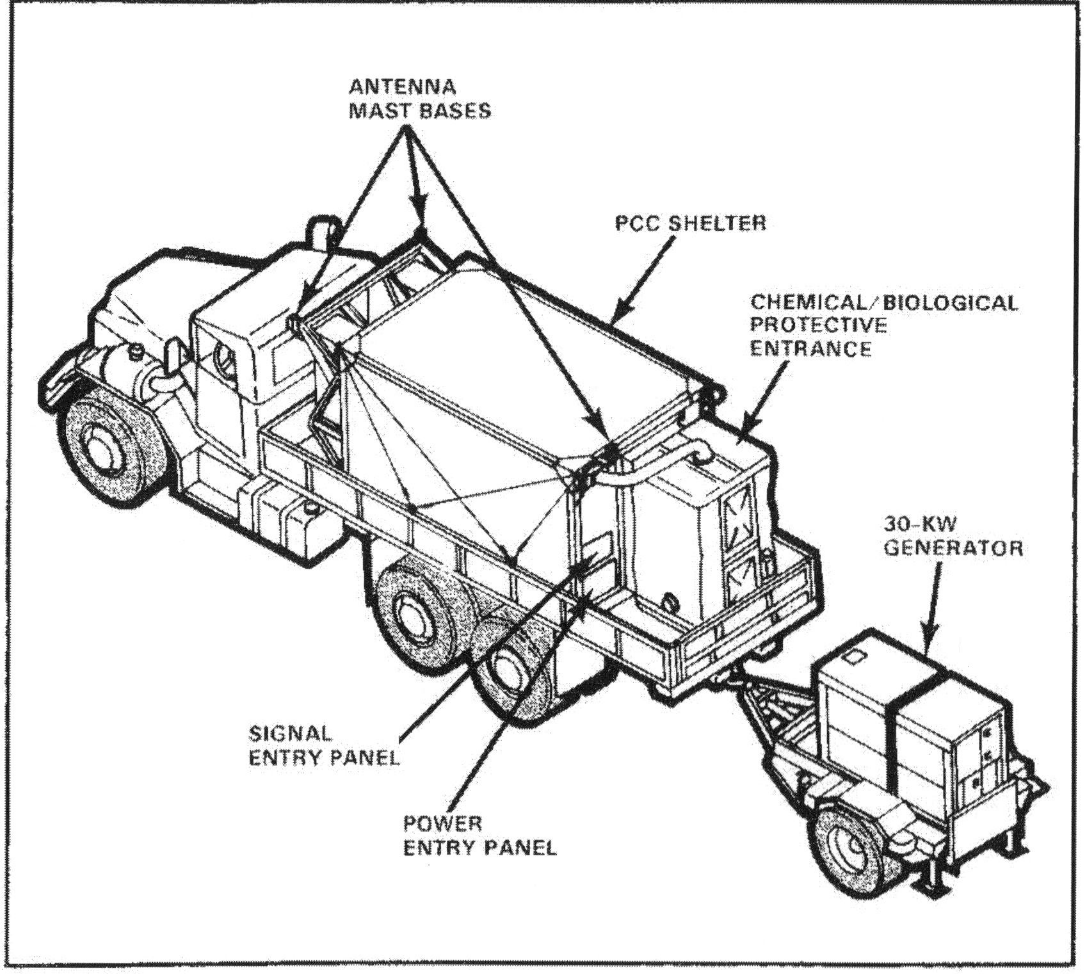

REFERENCE SCENE GENERATION FACILITY

There are two reference scene generation facilities (RSGF) per Pershing II battalion. The RSGF consists of equipment to generate target cartridges. It uses a digital data base provided by the Defense Mapping Agency (DMA) to produce magnetic tape cartridges of digital target scenes for entry into the missile through the GIEU. The RSGF is mounted on a 5-ton, long-wheelbase truck. It contains a computer, mass storage units, video displays, and a tape cassette unit. The power source for the RSGF is a trailer-mounted 30-kw generator.

CABLES

Data and power are distributed within a platoon area through a ground support cable network.

Two primary cables connected to the erector-launcher are required to launch the missile:

■ A 50-foot cable provides power from the 30-kw generator set to the erector-launcher through a power distribution box housed on the prime mover.

■ A 400-foot cable conducts all electronic signals between the PCC and the erector-launcher.

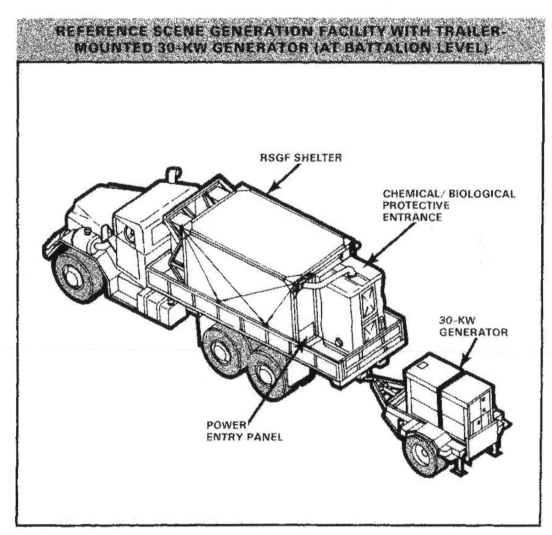

REFERENCE SCENE GENERATION FACILITY WITH TRAILER-MOUNTED 30-KW GENERATOR (AT BATTALION LEVEL)

RSGF SHELTER

CHEMICAL / BIOLOGICAL PROTECTIVE ENTRANCE

30-KW GENERATOR

POWER ENTRY PANEL

Additional cables include the following:

■ A 7-foot cable connecting the 30-kw generator set with the power distribution box.

■ A 150-foot AC power cable connecting a commercial or an alternate standby power source to the power distribution box. This cable connects power distribution boxes for distribution of standby power.

■ Three 50-foot ground cables with associated stakes, clamps, and rods.

■ Two short cables from the GIEU to the first stage, which provide power and electrical signals between the EL and the missile.

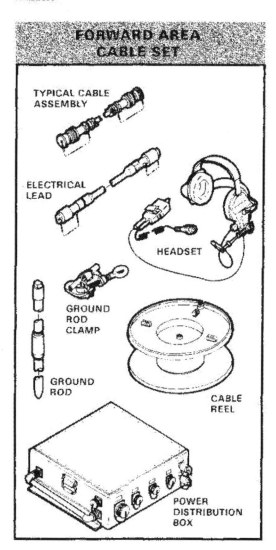

FORWARD AREA CABLE SET

TYPICAL CABLE ASSEMBLY

ELECTRICAL LEAD

HEADSET

GROUND ROD CLAMP

GROUND ROD

CABLE REEL

POWER DISTRIBUTION BOX

CONTAINERS

Containers are used to transport or store missile sections that are not assembled on an erector-launcher. All containers are top loading. The container covers are lifted by two top-mounted hoisting eyes. All missile section containers are designed so in-container tests can be performed through access ports without removal of the container covers.

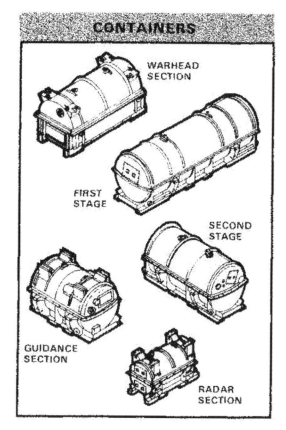

CONTAINERS

WARHEAD SECTION

FIRST STAGE

SECOND STAGE

GUIDANCE SECTION

RADAR SECTION

SLING SETS

The sling sets used in the PII firing battery consist of two- and four-leg slings, propulsion section hoisting beams, and a universal sling.

The *two-leg sling* is used to remove and replace all missile section container covers, including the warhead section container cover. It is also used to handle either of the propulsion section hoist beams, with or without the propulsion section attached.

The *four-leg sling* is used to handle missile section containers (loaded or empty) and the PCC and RSGF shelters. Its construction is similar to that of the two-leg sling.

The *propulsion section hoisting beams* are provided in two sizes. The larger hoisting beam is used to handle the first stage, and the smaller hoisting beam is used to handle the second stage. The hoisting beams are used to—

■ Lift the propulsion sections to or from their containers.

■ Position the sections on the launcher.

■ Align the sections for assembly operations.

The *universal sling* is used to lift either the G&C/A section, the warhead section, or the radar section. Three pairs of mounting holes are provided for lifting strap spacing to permit balanced lifting of the different sections.

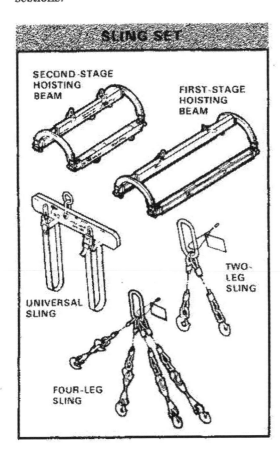

TARGET REFERENCE SCENE PRODUCTION

The reference scene generation facility is used to produce target cartridges. These cartridges, containing reference scenes and target data, are programed into the on-board computer of the missile. They are later compared with the live radar scan from the reentry vehicle. To produce target reference scenes, the RSGF must use information extracted from a target list and an operational data base.

OPERATIONAL DATA BASE

The operational data base (ODB) contains digitized elevation and topographic feature data stored on discs. These ODB discs are produced by the DMA and distributed to all Pershing II units.

TARGET CARTRIDGES

Target cartridges contain reference scenes and target data. Target data is programed into the missile by inserting the target cartridge into the GIEU on the erector-launcher. From the GIEU, the recorded target data is transferred to the on-board computer in the missile.

TARGET CARTRIDGE MANAGEMENT

The battalion targeting section, headed by the targeting officer, controls and distributes the cartridges as part of its tactical fire direction efforts. A target-to-cartridge assignment list received from the brigade headquarters specifies the arrangement of reference scenes for planned targets on each cartridge to be generated. Cartridges are labeled and become controlled documents. Reference local regulations for the identification, classification, security, and control of target cartridges.

Field Storage. During field operations, the platoon control central is the storage location for all cartridges in the platoon's possession. The PCC safe has one drawer with a special rack that can hold a platoon's complement of target cartridges in a readily accessible configuration. Cartridges are removed from this safe only during countdown operations, cartridge exchange, or garrision storage.

Countdown Operations. When a cartridge is moved from the PCC to the firing site, it must be transported by a person with a clearance equal to or higher than the security classification of the cartridge.

Cartridge Exchange. If cartridges must be exchanged for target change or maintenance reasons, classified material receipts will document the exchange. Defective cartridges found in firing unit countdown operations will be returned to the battalion targeting section for replacement. If the battalion targeting personnel can neither declassify the target data nor restore the cartridge to operational condition, the cartridge must be turned in for destruction as classified material. A replacement cartridge must be generated and fielded.

COUNTDOWN OPERATIONS

The Pershing II missile system can perform four types of countdowns: standard count, confidence count, quick count, and T-1 count. Each count accomplishes specific functions within the missile, ensuring accomplishment of fire missions. TM 9-1425-386-10-2 provides step-by-step procedures for performing countdowns.

STANDARD COUNT

The standard count provides the sequence of operations to launch a missile, beginning with the missile in the travel configuration on the erector-launcher. The purpose of the standard count is to perform all preflight checks and presets while preparing the missile for launch. Normally, the countdown terminates with missile lift-off.

CONFIDENCE COUNT

The confidence count is performed to verify equipment. It accomplishes the same sequence of preflight checks and presets as does the standard count. However, the confidence count ends when ALIGNMENT COMPLETE is indicated on the status display panel. At termination of the confidence count, the missile is placed in a standby or hot-hold condition. This allows the missile to undergo a quick count at a later time and be fired. A confidence count is performed periodically to verify system

reliability. If needed, a confidence count may be switched to a standard count and the missile launched.

QUICK COUNT

The quick count is used for firing a missile after a confidence count has been performed. It begins with the missile in a standby condition and bypasses several of the preflight checks and presets associated with a standard count. The quick count may end with missile lift-off or, as with a confidence or standard count, any time before launch sequence initiation. The system may be put in either a standby or a hot-hold condition.

T-1 COUNT

The T-1 count is a training and evaluation vehicle. It allows an individual to see all the missile functions that occur during a standard or quick count, including missile erection, but without firing the missile. This count may be used in the training and evaluation of missile crews as well as in the evaluation of the missile system.

TWO-STAGE FLIGHT SEQUENCE

FIRST-STAGE IGNITION

Milliseconds after ignition and lift-off, the missile begins to pitch, or tilt, toward the target at a predetermined rate. Initial thrust is provided by the first-stage rocket motor, which burns completely regardless of target range.

COAST PERIOD

After the first stage burns out, the missile enters a short coast period before the Pershing airborne computer (PAC) issues the separation signal.

FIRST-STAGE SEPARATION, SECOND-STAGE IGNITION

The first in-flight separation occurs between the first and second stages. A linear shaped charge in the second-stage aft skirt detonates, which causes the first stage to separate from the missile. The second stage then ignites, accelerating the remaining missile sections along the flight path.

SECOND-STAGE SEPARATION

When the values determined by the on-board computer indicate the missile is on the correct flight path and has reached the necessary velocity and range, an in-flight separation occurs. Explosive devices in the adapter detonate. This causes an in-flight separation between the second-stage propulsion section and the reentry vehicle. Simultaneously, the thrust termination ports in the forward end of the second-stage rocket motor are activated to terminate the second-stage thrust.

REENTRY AND TERMINAL GUIDANCE

The reentry vehicle, with its on-board terminal guidance system, provides guidance through the remaining trajectory to impact. Through the use of thruster assemblies, the reentry vehicle maintains attitude control outside the atmosphere. Air vanes provide control once the reentry vehicle reenters the atmosphere. The radar in the nose of the reentry vehicle is activated during the final portion of flight and scans the terrain in the region of the target area. The computer converts the radar image to a digital representation of the target area, compares this "live scene" to a previously stored reference, computes the adjustments necessary to hit the target, and applies those corrections using air vanes on the reentry vehicle. This "search, compare, correct" routine is repeated several times during the final phases of the trajectory. It provides the same accuracy regardless of range.

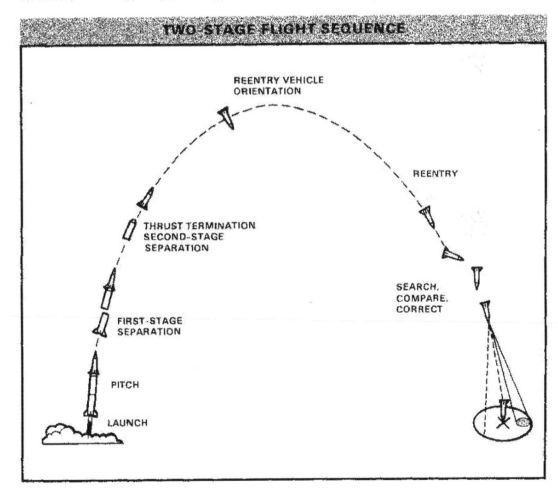

TWO-STAGE FLIGHT SEQUENCE

REENTRY VEHICLE ORIENTATION

REENTRY

THRUST TERMINATION
SECOND-STAGE
SEPARATION

SEARCH,
COMPARE,
CORRECT

FIRST-STAGE
SEPARATION

PITCH

LAUNCH

SINGLE-STAGE FLIGHT SEQUENCE

The single-stage flight sequence is very similar to the two-stage flight sequence.

FIRST-STAGE IGNITION

First-stage ignition occurs as in the two-stage flight sequence. After lift-off, the missile maneuvers to a precomputed firing azimuth. The missile continues in flight under pitch and yaw control of a swivel nozzle and roll control of the first-stage air vanes. The first-stage motor burns until the proper velocity is achieved to allow free-fall onto the target.

FIRST-STAGE SEPARATION

There is no coast period in the single-stage flight sequence. Once the proper velocity is reached, the on-board computer issues a cutoff signal. At this time, the thrust reversal system (thrust termination ports) activates, the RV separation system activates, and the RV separates from the first stage at the adapter section.

REENTRY AND TERMINAL GUIDANCE

The reentry phases are identical for single-stage and two-stage flight sequences.

CHAPTER 2
TASKS, ORGANIZATION, AND KEY PERSONNEL

The mission of the field artillery (FA) is to destroy, neutralize, or suppress the enemy by cannon, rocket, and missile fire and to help integrate fire support into combined arms operations. Pershing II gives the theater commander the ability to strike targets accurately at great range. Because of the confidence count and standby capabilities of the system, Pershing II units can be ready over long periods to react quickly to fire orders against planned targets. Pershing-peculiar organizations and duties are discussed in this chapter.

DEPLOYMENT TO FIELD LOCATIONS

To continue the deterrence role and enhance the system's survivability during periods of increased international tension or war, Pershing II units will deploy to field locations.

The battalion commander gives each firing battery an area in which to position its elements. Normally, the battalion area will be quite large to accommodate the movement of the various elements.

A firing battery occupies three distinct platoon positions separated by at least 3 to 10 km. Each light platoon position contains the firing platoon itself and enough communications, operations, and maintenance support for semiautonomous operations. The platoon leader is the OIC of a light platoon position. A heavy platoon position contains the third firing platoon and the bulk of the communications, operations, maintenance, and headquarters elements. The battery commander (BC) is the OIC of the heavy platoon position and, as in all command functions, is responsible for mission accomplishment by all elements of the battery.

Movement to initial field positions is directed by an emergency action message received by the battalion operations center (BOC). The battalion commander, through the BOC, coordinates and directs the batteries' movements in accordance with SOPs. Normally, a battery will move to the field from garrison in four serials. These include the two light firing platoons, the firing elment of the heavy platoon, and the headquarters/support element of the heavy platoon.

FIRING BATTERY TASKS

A firing battery may be given any one of the following four tasks to perform in the field:

■ Assume quick reaction alert (QRA) target coverage immediately upon arriving at the position (immediate coverage). A missile is considered to be in QRA status when it has been prepared for a quick count and has been placed in standby.

■ Prepare to assume target coverage at a later date (delayed coverage).

■ Assume a hide configuration (maximize survivability) and await future taskings (no coverage).

■ Launch a single missile (single taraget).

IMMEDIATE COVERAGE

The firing position facilitates the task of immediately assuming QRA target coverage. The firing element completely emplaces the missiles and performs confidence counts in preparation for firing. All of the position's communications assets are operative, and the unit enters all appropriate communications nets. While the primary task of a unit occupying a firing position is to be prepared to fire its missiles with minimum reaction time, the defense of the unit must not be forgotten. The QRA target coverage must be assumed as quickly as possible. Vehicles and equipment are positioned to minimize the possibility of detection from the air or from the ground. Communications must be minimized to avoid detection by radio direction finding.

DELAYED COVERAGE

The unit with the task of preparing to assume target coverage some time after deploying occupies a silent firing position. The silent firing position is configured basically the same as the firing position. The primary concern, however, is to avoid detection for as long as possible. Missiles are emplaced, but no countdowns are performed. Whenever possible, radio silence is observed. Normally, generators are not operated, and vehicle traffic should be kept to a minimum. Only when the unit is directed to assume target coverage will any countdowns be performed. Survivability is the key.

NO COVERAGE

The hide position is the primary survivability configuration for Pershing II. A unit that is not expected to assume target coverage in the near future will be directed to occupy a hide position. As the name implies, the unit hides from the enemy. A tight vehicle configuration is most commonly used. Erector-launchers are not emplaced. Radio silence is observed unless the hide task is changed or cancelled. The unit may be tasked to fire single missiles from an external firing point. The equipment necessary to fire a single missile must be positioned to facilitate displacement and firing.

SINGLE TARGET

Single missiles may be fired from a unit's current position or, if time is available and the tactical situation permits, from an external firing point (EFP). An EFP should be located at least 5 km from the unit's principal position. It should provide adequate cover and concealment from enemy detection. The purpose of the EFP is to avoid compromising the main platoon position and to enhance survivability. Care should be taken to avoid compromising the location of the EFP or the main platoon position during organization of, or movement to, the external firing point.

ORGANIZATION

The Pershing II firing battery is organized with a battery headquarters, an operations/communications platoon, three firing platoons, and a support platoon.

The *battery headquarters* contains the personnel and equipment to provide:

■ Command and control.

■ Personnel services.

■ Supply support.

■ Nuclear surety administration.

■ Nuclear, biological, and chemical (NBC) defense and detection expertise.

■ Food service support.

The *operations/communications platoon* provides the personnel and equipment required to direct missile operations, maintain the battery crypto account, and direct the employment of communications assets.

The *three firing platoons* deploy, maintain, secure, and simultaneously fire three Pershing missiles. Personnel and equipment are provided to transport and assemble the missiles, input targeting data, operate the platoon control central (which controls missile launching), launch missiles, maintain missiles, secure nuclear weapons, and secure the position area. Administration,

supply, food service, communications, and maintenance support are provided by the battery.

The *support platoon* manages all aspects of unit maintenance and directly supervises the distribution of petroleum, oils and lubricants (POL). The platoon has two sections—automotive maintenance and missile maintenance. During field operations, personnel from the maintenance sections are deployed with the firing platoons.

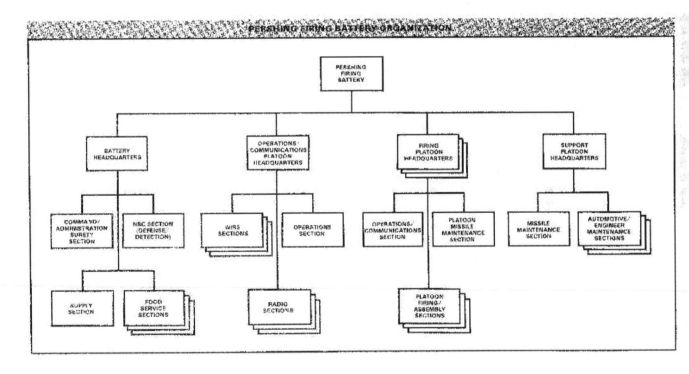

BATTERY COMMANDER

The battery commander (MAJ, 13C) is ultimately responsible for everything the battery does or fails to do. He ensures mission accomplishment. The battery commander must respond to calls for fire directly from the theater combined headquarters, joint headquarters, and Army headquarters under extraordinary rules and procedures. He is directly responsible for the movement, employment, and firing of his missiles, which are in three separate firing locations. Some of the battery commander's responsibilities follow:

■ Reconnoiter and select battery areas that afford mission accomplishment while providing for survivability.

■ Plan specific actions to enhance the battery's survivability.

■ Plan unit marches and movements.

■ Supervise the security, preparation, and delivery of nuclear weapons.

■ Keep the battalion commander and battery personnel informed.

■ Establish and maintain communications and electronics security.

■ Plan for battery supply and maintenance.

■ Manage the battery's personnel reliability program (PRP).

■ Ensure that all battery personnel are trained for combat.

■ Ensure that all battery equipment is maintained in accordance with applicable technical publications.

EXECUTIVE OFFICER

The executive officer (XO) (CPT, 13C) is the battery commander's principal assistant and acts for the commander in his absence. He helps the commander command and control an organization spread over three locations. He coordinates and manages the battery's unit maintenance. He also supervises a shift in the battery control central (BCC) to facilitate 24-hour sustained operations. His duties include the following:

■ Ensure that timely and accurate fires are delivered by all firing platoons.

■ Ensure that before-, during-, and after-operation maintenance is performed in accordance with exact procedures stated in technical manuals.

■ Ensure continual position area improvement.

■ Ensure that intrabattery communications are maintained.

■ Ensure that all safety procedures are followed.

■ Perform/supervise destruction procedures to prevent enemy use of battery equipment.

■ Supervise nuclear weapon release procedures.

■ Perform courier officer/convoy commander duties.

OPERATIONS OFFICER

The operations officer (CPT, 13C), assisted by the operations sergeant (E7, 15E), performs the same functions for the battery as the S2, S3, and communications-electronics staff officer (CESO) do for the battalion. As a key decision maker during normal firing operations, he establishes and maintains coverage of targets assigned to the battery. He is the platoon leader for the operations/communications platoon. His other responsibilities include the following:

■ Maintain status of nine missiles in three separate locations.

■ Assign targets and target data to include a real-time inventory of reference scenes.

■ Maintain control over PAL and emergency message authentication system (EMAS) material.

■ Coordinate battery ammunition resupply and special weapons security.

■ Coordinate security with supporting infantry personnel.

■ Perform/supervise destruction procedures to prevent enemy use of battery equipment.

■ Perform/supervise nuclear weapon release procedures.

■ Perform training officer functions.

■ Control all classified documents and communications security (COMSEC) materials.

■ Provide for continuous BCC operations.

■ Ensure proper use of wire and radio nets.

■ Ensure that accurate records of missions are maintained.

■ Perform courier officer/convoy commander duties.

■ Work shifts as BCC officer in charge.

FIRING PLATOON LEADER

The firing platoon leader (CPT, 13C) performs the same functions for the firing platoon, except administering the Uniform Code of Military Justice (UCMJ), as the BC does for the battery. In addition to being responsible for the platoon's performance, his responsibilities include the following:

■ Select suitable positions for the platoon to occupy.

■ Thoroughly train the advance party.

■ Perform courier officer/convoy commander duties.

■ Supervise special weapons operations.

■ Perform/supervise nuclear weapons release procedures.

■ Derive pace data to launch points for use in preparing manual data entries (MDE).

■ Verify manual data entries.

■ Thoroughly train the missile crews.

■ Manage platoon equipment maintenance.

■ Employ and control platoon communications assets.

■ Employ and control platoon security assets.

■ Work shift in the platoon control central.

■ Ensure continual position improvement.

■ Perform destruction to prevent enemy use of platoon equipment.

FIRE CONTROL OFFICER

The fire control officer (FCO) (LT, 13C) is the principal assistant to the firing platoon leader. While his main place of duty is in the PCC, the FCO must be prepared to act for the platoon leader in his absence. His duties include the following:

■ Work shift in the platoon control central.

■ Supervise PCC operations including PAL, safe/arm, countdown, and launch procedures.

■ Prepare manual data entries.

■ Compute launch data and verify target cartridges by Pershing identification (PID) number.

■ Secure and control classified documents within the platoon.

■ Perform/supervise nuclear weapons release procedures.

■ Develop and conduct the training program for PCC personnel.

■ Move the platoon to the next location on order.

■ Ensure continuous PCC operations.

■ Ensure proper use of wire and radio nets.

■ Ensure that accurate PCC records of missions are maintained.

■ Perform courier officer/convoy commander duties.

■ Perform destruction to prevent enemy use of platoon equipment.

SUPPORT PLATOON LEADER

The support platoon leader (LT, 13C), assisted by the senior maintenance supervisor (E7, 63B), supervises and trains the support platoon. His duties include the following:

■ Advise the BC on the formation of his maintenance program.

■ Execute the BC's maintenance program.

■ Advise the BC and XO on the maintenance status of battery equipment, and provide information for readiness reports.

■ Manage the battery's prescribed load list (PLL) and POL resources.

■ Perform courier officer/convoy commander duties.

MISSILE MAINTENANCE TECHNICIAN

The missile maintenance technician (WO, 214EO) is responsible for the supervision of all scheduled and corrective missile maintenance and liaison with the forward support company elements supporting his platoon.

His duties include the following:

■ Maintain platoon missile equipment, maintenance records, and status reports.

■ Coordinate maintenance beyond his capabilities with the support platoon leader and executive officer.

■ Help the platoon leader select a position.

■ Help the platoon leader during special weapons operations.

■ Give technical assistance/advice during missile operations (emplacement, counting, firing, march order, assembly).

■ Perform courier officer/convoy commander duties.

■ Perform destruction to prevent enemy use of platoon equipment.

FIRST SERGEANT

The first sergeant (1SG) (E8, 13Y) is the battery commander's principal enlisted assistant. His responsibilities include the following:

■ Ensure that enlisted supervisors are adequately trained.

■ Ensure that administrative and personnel actions within the battery are handled efficiently.

■ Maintain status of all assigned enlisted personnel.

■ Train enlisted members of the heavy position advance party.

■ Assemble the heavy position advance party.

■ Help reconnoiter and select battery areas.

■ Establish the track plan for occupation of the heavy position.

■ Supervise occupation of and position support vehicles in the heavy position.

■ Develop and brief the heavy position defense plan.

■ Designate and rehearse the heavy position reaction force.

■ Detail personnel to perform support functions in the heavy position (perimeter defense, sanitation, and kitchen police).

■ Coordinate administrative and logistical support.

■ Ensure that the light platoons have adequate defense plans, logistical support, and so forth.

■ Perform duties as courier officer/convoy commander.

FIRING PLATOON SERGEANT

The firing platoon sergeant (E7, 15E) is the firing platoon leader's principal enlisted assistant. Just as the platoon leader functions as the battery commander, the platoon sergeant functions as the first sergeant. His responsibilities include the following:

■ Ensure that all enlisted personnel of the platoon are adequately trained.

■ Help reconnoiter and select platoon positions.

■ Supervise the security sweep of the platoon position.

■ Establish a track plan for occupation of the platoon position.

■ Supervise occupation of and position support vehicles.

■ Develop and brief the platoon position defense plan.

■ Designate and rehearse the reaction force.

■ Detail personnel to perform support functions.

■ Supervise the development of range cards for crew-served weapons.

■ Advise the chief of section.

■ Perform destruction to prevent enemy use of platoon equipment.

■ Perform duties as courier officer/convoy commander.

FIRING PLATOON CHIEF OF SECTION

The firing platoon chief of section (E6, 15E) is the platoon sergeant's principal assistant. He deals directly with missile operations. His duties include the following:

■ Ensure that all personnel in his section are properly trained.

■ Ensure readiness of his missiles.

■ Ensure that during countdown/firing, all crew duties are performed.

■ Ensure that all prefiring checks are made.

■ Ensure that all section equipment is on hand and properly maintained.

■ Ensure that during missile assembly/disassembly, all crew duties are performed.

■ Prepare range cards for crew-served weapons.

■ Supervise the preparation of fighting positions for both individual and crew-served weapons.

■ Be familiar with the position defense plan.

■ Implement his portion of the defense plan.

■ Perform duties as safety NCO.

■ Perform destruction to prevent enemy use of platoon equipment.

■ Perform duties as special weapons custodial agent.

CHAPTER 3
RECONNAISSANCE, SELECTION, AND OCCUPATION OF POSITION

On the battlefield, a sophisticated enemy can locate and engage a battery/platoon in various ways. To survive, it may have to move frequently. Frequent movement, however, reduces responsiveness and necessitates greater reliance on other units to assume the mission during displacement. To minimize movement time, all key personnel must be able to perform the reconnaissance, selection, organization, march, and occupation tasks quickly and efficiently. The keys to successful reconnaissance, selection, and occupation of position (RSOP) are discipline and team effort, which come from frequent and effective training.

The headquarters controlling the movement of the battery directs the essential elements of the movement—when, where, and how. The battery commander/platoon leader must anticipate movement. He must plan in advance for displacement to new, alternate, or supplementary positions. The BC should advise the controlling headquarters of any factors to be considered in determining the who, when, where, and how of the movement. The BC must provide for coordination of survey support to platoons. He must consider the factors of mission, enemy, terrain, troops available, and time (METT-T) when selecting positions for his area of operations.

```
┌─────────────────────────────────────────────┐
│            FACTORS OF METT-T                  │
│                                               │
│  MISSION:      Is the unit to operate in the  │
│                hide, silent, or firing mode?  │
│                                               │
│  ENEMY:        What is the current threat?    │
│                                               │
│  TERRAIN:      Can the unit communicate       │
│                with higher and lower          │
│                elements?                      │
│                                               │
│                Does the terrain support signal│
│                security?                      │
│                                               │
│                Does the terrain offer cover   │
│                and concealment?               │
│                                               │
│                Is the terrain defensible?     │
│                                               │
│                Will the terrain support       │
│                movement of the unit's         │
│                equipment under all weather    │
│                conditions?                    │
│                                               │
│  TROOPS        Does the unit have enough      │
│  AVAILABLE:    troops to defend the position  │
│                and remain mission-capable?    │
│                                               │
│  TIME:         How much time does the unit    │
│                have to accomplish the         │
│                mission?                       │
└─────────────────────────────────────────────┘
```

SECTION I
RECONNAISSANCE AND THE ADVANCE PARTY

DEFINITION

Reconnaissance is the examination of terrain to determine its suitability for accomplishment of the battery mission.

RECEIPT OF ORDER

The battery commander/platoon leader may receive displacement orders in varied formats, ranging from a five-paragraph operation order to a simple authenticated radio message. He is given, or selects himself, the general location of his new position, the time to depart and/or be in the new position, and the routes to be used.

METHODS OF RECONNAISSANCE

MAP RECONNAISSANCE

Once the battery commander has been assigned his "goose egg," he identifies all potential positions within his allocated area and routes to those positions. He then assigns areas within which the platoon leaders perform detailed reconnaissance. A line of sight profile sketch may be included as part of the map reconnaissance to preclude problems with FM communications between positions.

AIR RECONNAISSANCE

Air reconnaissance is a useful supplement to the map reconnaissance. It decreases the time required to assess a large battery position area. Care should be taken to avoid disclosure of position areas to enemy covert forces. When performing air reconnaissance, a commander should look for an area that provides a maximum number of launch and hide positions while avoiding open areas. He should concentrate on wooded or urban areas which are defensible and which permit maximum concealment of firing positions.

GROUND RECONNAISSANCE

A ground reconnaissance normally follows a map reconnaissance. This is the best method of determining the suitability of routes to be traveled and positions to be occupied. The actual conditions of routes and terrain patterns within the proposed area are seen. Before leaving his platoon, the platoon leader must give key information to his second in command, to include the following:

■ Location of the new position area.

■ Possible routes.

■ Mission and enemy situation.

■ Any peculiar aspects of positions and/or routes if known.

He goes to each of the positions identified through map or air reconnaissance. A survey team may go with him. While moving, he verifies the suitability of the routes. For example, he notes bridge and road classifications, location of obstacles, and likely ambush sites. After the reconnaissance, he should return by an alternate route to verify its suitability for convoys if needed. Planning the ground reconnaissance must include measures to avoid detection and location of future positions by the enemy.

> Note. Each method of reconnaissance offers the BC/platoon leader a different, but complementary, perspective. At times, all three methods may be used. In most instances, the BC/platoon leader performs a map reconnaissance, selects a tentative route, and then makes a ground reconnaissance.

PLANNING THE RECONNAISSANCE

In a fast-moving tactical situation, the time available for planning the reconnais-sance, assembling the advance party, and conducting the RSOP may be only a few minutes. Before departure, the commander must consider the primary and alternate routes and distances to the new position area. He must give the XO/FCO that information.

The organization of the advance party must be tailored. The platoon leader/enlisted assistant takes *only* route guides and the personnel and equipment necessary to prepare the position for occupation. These personnel also provide their own defense and initial defense of the new position area.

A standard nucleus of advance party personnel should be established. The equipment required to prepare a new position should be preloaded or identified and kept so that it can be located and loaded without delay.

ASSEMBLING THE ADVANCE PARTY

For either a deliberate or a hasty occupation, a prearranged signal or procedures should be used to alert and assemble the advance party. The signal should be specified in the unit SOP, which should also list the advance party personnel, equipment, vehicles, and place of assembly.

ORGANIZATION OF THE ADVANCE PARTY

The makeup of the party is determined by the platoon leader and platoon sergeant. It is based on the tactical situation and assets available.

The advance party is preceded to the position area by the supporting infantry contingent. The infantry reconnoiters the selected route to ensure it is free of Threat forces. To avoid compromising actual positions, the infantry does not enter the position. Rather, it performs a security sweep around the position, sets up listening posts (LP)/observation posts (OP), and begins patrolling. Communications must be maintained between the infantry, the advance party, and the main body.

The size of the advance party must be kept to a minimum to avoid detection. Normally, platoon leaders lead the advance parties that organize the new platoon positions. During the organization phase, adequate security should be maintained to protect against a small, squad-size force. Positive measures must be taken to conceal the location of the unit and to avoid detection.

If NBC warfare has been initiated, an NBC survey monitoring team should go with the advance party to check for NBC contamination. Upon arrival at the position, the advance party should be in the NBC protective posture dictated by the situation. First, the survey monitor team takes environmental samples of the entrance to the area. The results of the sampling are brought to the officer in charge. He determines whether the position is tenable and if any change to the NBC mission-oriented protective posture (MOPP) is necessary. If the position is tenable, the security team sweeps the area and establishes a defensive perimeter.

MAKING THE RECONNAISSANCE

Before departing, the platoon leader must brief key personnel and advance party members.

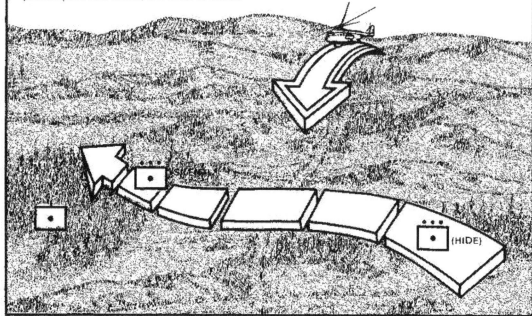

PLATOON LEADER'S BRIEFING

1. SITUATION

   a. *Enemy Situation.* Rear area activity, major avenues of approach, air activity, and potential ambush sites.

   b. *Friendly Situation.* Changes in tactical mission and location of adjacent units.

2. MISSION

   Changes in mission of the platoon, if any (for example, targeting, assignments and coverage, and generation levels).

3. EXECUTION

   a. *Concept of Operation.* General location of platoon positions, routes, and order of march.

   b. *Coordinating Instructions.* Location of start point, release point, and start point time if known/used.

   c. *Mission-Oriented Protective Posture Status.*

4. ADMINISTRATION AND LOGISTICS

   When and where to feed personnel, and priority for maintenance and recovery.

5. COMMAND AND SIGNAL

   a. *Command.* Changes in location of battalion/battery command post.

   b. *Signal.* Movement radio frequency and net control restrictions.

After being briefed by the platoon leader, the fire control officer/platoon sergeant should brief his remaining key personnel on the following:

- Tactical situation.
- Routes to be used.
- Any anticipated problems.
- Movement time, if known.

After making a map reconnaissance, completing his planning, and briefing necessary personnel, the platoon leader is ready to proceed to the new location. The primary purpose of the route reconnaissance is to verify the suitability of the primary route. The platoon leader also checks:

- Alternate routes (time permitting).
- Road conditions and bridge classifications.
- Cover and concealment.
- Location of obstacles.
- Likely ambush sites.
- Time required.
- Distance.
- Likely position areas along the route.

SECTION II
SELECTION OF POSITION

BASIC TYPES OF POSITIONS

The platoon leader must select three basic types of positions—primary, alternate, and supplementary. Positions are further identified as heavy or light, depending on the units occupying them.

PRIMARY POSITION

The primary position is one from which a platoon intends to accomplish its assigned mission.

ALTERNATE POSITION

The alternate position is the one to which the entire platoon moves if its primary position becomes untenable. Since the platoon will continue its mission from the alternate position, it must meet the same requirements as the primary position and should be far enough away that the unit can escape the effects of enemy indirect fire on the primary position. It should be reconnoitered and prepared for occupation as time permits. Each section chief must know the route to the alternate position, because movement to that position may be by section.

SUPPLEMENTARY POSITION

A supplementary position, such as an external firing point, is selected for accomplishment of a specific mission.

HEAVY AND LIGHT POSITIONS

The battery headquarters is collocated with a firing platoon in the heavy platoon position. The battery headquarters should be located to provide the best communications with the subordinate firing platoons and battalion headquarters. Each of the other two firing platoons is located in a light platoon position. A consideration in light platoon area selection is that fewer resources are available for defense. To enhance survivability, from 3 to 10 km between firing platoon positions is desired. Exact distance depends on the expected threat.

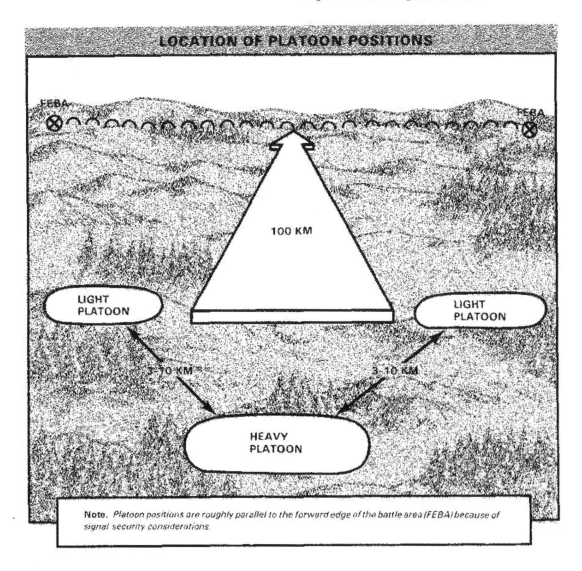

LOCATION OF PLATOON POSITIONS

Note. *Platoon positions are roughly parallel to the forward edge of the battle area (FEBA) because of signal security considerations.*

SPECIFIC TYPES OF POSITIONS

In addition to the three types of positions mentioned before, the platoon leader must select positions that support the tasks assigned. These are the firing, silent firing, and hide positions; an external firing point; and, if required, an assembly area.

FIRING POSITION

Accomplishing the mission of firing missiles must take precedence in selection of a firing position. While overhead cover is essential for survival, it must not interfere with a missile during lift-off. Therefore, at least a 10-foot hole must be in the overhead cover of each launch point. The position must be large enough that ancillary equipment can be placed out of the blast zones, yet small enough to defend. An established road network that can support the weight of an erector-launcher with missile and that affords separate entry and exit routes should be used when possible. Missiles must be positioned within the constraints specified by the appropriate technical manuals. Ancillary vehicles are normally positioned to protect the missiles from direct fire and ground observation from outside the position.

SILENT FIRING POSITION

The silent firing position increases survivability while affording a minimum response time for a fire mission. The RSOP considerations are the same as for a firing position. The equipment is emplaced for firing; however, no power is generated. Passive defense measures are strictly enforced, as for the hide position.

HIDE POSITION

When a platoon is not required to assume target coverage, survival is its primary task. All measures of passive defense must be strictly followed, as prescribed in chapter 4. The RSOP considerations for successful hide operations are as follows:

■ Missile equipment is not positioned for firing.

■ Equipment is positioned to reduce defense requirements.

■ An exclusion area is established for any nuclear weapons present.

■ A restricted area for the BCC/PCC, single sideband (SSB) radios, and tactical satellite (TACSAT) must be planned.

■ The hide position should be near selected firing positions. That would minimize the time required to move and begin firing operations.

■ Positioning should be planned so missile equipment can be easily moved out of the position without the entire platoon having to move.

■ There should be enough room between vehicles that if one vehicle becomes inoperable, others can get around it.

■ Equipment must be concealed from the sides as well as from overhead with enough cover to prevent detection by observers and side-looking airborne radar.

The best hide position allows expansion to a silent firing position or a firing position without moving the platoon to a different area.

EXTERNAL FIRING POINT

An external firing point is a position to which one missile is taken and prepared for launch. Covert operations are absolutely essential. The only equipment taken is that necessary to launch the missile (including communications and operations equipment), provide security, and communicate. An advance party is not needed in most cases. Because of the minimal equipment, the position is much smaller than that of a firing platoon. Noise and light discipline is an absolute must during all aspects of EFP operations.

ASSEMBLY AREA

After initial rounds are fired, battery assets may have to be consolidated for continued operations. The assembly area is designed to facilitate missile and warhead resupply as well as missile assembly operations before the platoons or sections of platoons occupy other firing positions.

Many of the criteria for the selection and occupation of firing positions apply to assembly areas. The major exception is that the mission to be performed normally does not include firing missiles. A long, wide, level, easily trafficable area must be selected to ease resupply and missile assembly operations and to hold the remaining assets of the entire battery. The area also must provide necessary concealment.

The size of the assembly area depends on the amount of equipment it will contain. For a complete firing battery, the position must be extremely large. Extra fighting positions on the defensive perimeter may be necessary.

COMPARISON OF POSITIONS

	HIDE POSITION	SILENT FIRING POSITION	FIRING POSITION	EXTERNAL FIRING POINT	ASSEMBLY AREA
Missiles emplaced	No	Yes	Yes	Yes	No
Confidence counts	No	No	Yes	No	No
Standard counts	No	No	No	Yes	No
Communications assets	FM only	FM only	All	All	FM only
Messing	Cold	Cold	Hot	Cold	Cold
Overhead cover	Full	Holes	Holes	Holes	Full
Generators	1.5 kw	1.5 kw	All	All	1.5 kw
Camouflage	Natural and man-made	Natural and man-made	Natural and man-made	Natural	Natural and man-made

SECTION III
ORGANIZATION OF POSITION

TASKS IN ORGANIZATION

Organization of the platoon position includes:

■ Designating locations of vehicles, facilities, and equipment.

■ Briefing vehicle guides.

■ Preparing the track plan.

■ Designating vehicle order of march for the main body.

■ Designating chemical agent sampling locations.

■ Preparing the defense plan.

■ Preparing a plan for internal wire communications. When time permits, wire should be laid before the main body arrives.

LAUNCH POINT IDENTIFICATION

A firing position must allow for safe and unobstructed launch of all missiles. Availability of an adequate launch area drives the selection of the position. (A launch area in a firing position normally contains three missiles on erector-launchers.) The major considerations in launch area selection include the following:

■ Launch paths—Overhead cover must not obstruct missile firing.

■ Blast zones—The area must be large enough to allow equipment emplacement away from the missile blast. See TM 9-1425-386-10-2/1. Blast effects, not only on personnel but also on low-lying vegetation, should be considered.

■ Terrain slope tolerance—A slope of 6° or less is required for erector-launchers.

■ Erector-launchers—The erector-launchers must be concealed from air and ground observation. Missiles should be protected from direct engagement by small-arms fire from outside the perimeter.

■ Emplacement positions for erector-launchers must be able to support the weight of the vehicle and allow erection of the missile.

■ Distance between launch points cannot exceed interconnecting cable lengths.

FACILITIES AND EQUIPMENT

The current situation dictates the need for tactical dispersion of equipment, for natural cover and concealment, and for other passive defense measures in position selection. Capabilities and limitations of communications equipment must be considered in siting the platoon headquarters. The BCC/PCC and defense control center (DCC) positions must be selected to facilitate radio communications as well as landline communications to the defensive perimeter and the launch area. The position must facilitate missile section and RV loading and mating operations. There should be room to position vehicles delivering sections and RVs as well as to reposition the materiel-handling crane mounted on the prime mover. The position should facilitate generator refueling as required.

The position must facilitate both active and passive defense measures. A defensible position should—

■ Deny enemy observation during occupation.

■ Provide natural concealment.

■ Provide multiple entry and exit routes.

■ Permit the unit to mask communications emitters behind hill masses.

■ Provide adequate fields of fire and visibility to properly defend the position.

■ Permit emplacement of all equipment in a small area so that a minimum number of fighting positions can adequately support a defensive perimeter.

TRACK PLAN AND VEHICLE ORDER OF MARCH

The area must be such that vehicles can get in and out of it. A Pershing II platoon has many large, wide, and heavy vehicles. Tree density, softness of ground, and slope of terrain must be considered. Missiles on erector-launchers will probably have to stay on hard-surface or semihard-surface roads because of their extreme weight. There should be sufficient drainage in the area that a platoon will not get bogged down during a heavy rain. The track plan should clearly show how vehicles will move into and out of the position. It should also identify the main body order of march, the convoy release point, and the vehicle guide pickup point. The vehicle order of march should be given to the main body before movement, and it should facilitate occupation.

DEFENSE PLAN

The defense plan should be a sketch of the defensive perimeter, observation/listening posts, patrol areas, minefields, and early warning measures. It must provide for 360° defense of the perimeter as well as defense in depth. Missiles must be protected from small-arms fire from outside the perimeter. Auxiliary vehicles may be used for protection of missiles and then be moved out of blast zones before launch. Chapter 4 presents detailed defensive considerations.

INTERNAL WIRE PLAN

The wire plan should identify each TA-312 telephone and termination point within each heavy and light platoon area. The perimeter positions will have wire communications with the DCC switchboard. Wire communications will be maintained between the DCC and the BCC/PCC, the mess facility, observation and listening posts (OPs/LPs will have to rely on FM communications during initial emplacement), and the exclusion area gate (X-gate). The BCC/DCC must keep direct wire communications with all radio facilities. If the reaction forces are not at the DCC, communications must be maintained from the DCC to these forces.

VEHICLE GUIDES

Guides must know where each vehicle and facility will be located as specified by the track plan. They must also know in which direction to point the vehicle to facilitate its departure. Guides must walk the route from the pickup point to the vehicle/facility location and ensure that the route is unobstructed. When moving vehicles into the position area, guides begin controlling vehicle movement at the pickup point. Vehicles must be moved off the road as soon as possible to avoid detection. If there are not enough guides for each vehicle, guides should be located for maximum control and to help assistant drivers quickly position vehicles and equipment.

SECTION IV
TACTICAL MARCHES

MOVEMENT METHODS

Normally, the firing battery moves by platoon. The movement order for the main body includes start point (SP) and release point (RP) times as well as the order of march. A vehicle from the advance party meets the main body at the release point and leads the convoy to the pickup point. Vehicle guides control vehicle movement from the pickup point to the individual vehicle locations. (Vehicles carrying warheads require point guards at this time.) The position should be quickly occupied as specified in the track plan. Units must be drilled extensively on all movement methods to minimize the time they are unable to provide fire support. The loading of equipment on vehicles and trailers begins upon receipt of a warning order to move. Consistent with the mission, communications equipment is switched from the static to the mobile mode of operation. Missiles and related equipment should be prepared for transport as soon as possible.

There are two primary methods of moving a Pershing unit in a tactical configuration—open column and closed column. Each method has it specific advantages and disadvantages. The battery commander/platoon leader decides which method is best.

OPEN COLUMN

The open column is used for daylight movements whenever there is an adequate road network that is not overcrowded, when enemy detection is not likely, when time is an important factor, and when there is considerable travel distance involved. The vehicle interval in an open column is generally 100 meters.

Advantages of this method are:

■ Speed (the fastest method of march).

■ Reduced driver fatigue.

■ Improved vision on dusty roads.

■ Ease in passing individual vehicles.

■ Ease in dispersing vehicles as a passive defense measure against an air attack.

■ Less chance of the entire unit being ambushed.

Disadvantages of this method are:

■ Greater column length requires more road space.

■ Other traffic often becomes interspersed in the column.

■ Communication within the column is complicated.

CLOSED COLUMN

For closed column movement, the vehicle interval is less than 100 meters. At night, each driver can watch the cat's-eyes of the blackout markers on the vehicle in front of him and keep an interval of 20 to 50 meters. If the driver sees two marker lights, the interval is too great. If he sees eight marker lights, he is too close. If he sees four marker lights, he is maintaining the proper interval. During daylight, closed column is used when there is a need for maximum command and control;

for example, during periods of limited visibility or when moving through built-up or congested areas.

Advantages of this method are:

■ Simplicity of command and control.

■ Reduced column length.

■ Concentration of defensive firepower.

Disadvantages of this method are:

■ Column is vulnerable to enemy observation and attack.

■ Strength and nature of the column are quickly apparent to enemy observers.

■ Convoy speed is reduced.

■ Driver fatigue increases.

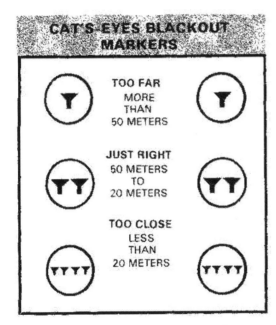

LOADING PLAN/LIST

A loading plan/list is particularly important in sustained combat, and each vehicle should have one. The loading plan should be recorded and graphically illustrated. The loading list need not be graphically illustrated. The plan/list helps ensure the unit will close on the new position with all its equipment. Personnel responsible for preparing loading plans/lists should consider the mission, personnel, SOP, and equipment of the battery. Equipment needed

first during occupation should be loaded last. Steps in preparing the loading plan/list include:

■ Examine the modified tables of organization and equipment (MTOE) to determine the personnel, equipment, and vehicles authorized for each section.

■ Examine all tables of distribution and allowances (TDA) property that must be transported by the battery. This equipment should be carried by the section that uses it.

■ List the personnel and equipment to be carried in each vehicle. Equipment should be loaded the same way for each move to aid in identification under blackout conditions. Technical manual guidance, if available, should be followed.

■ Practice loading, and adjust the plan if necessary.

■ Establish a list of items that must be removed from the vehicle and carried forward if the vehicle becomes disabled during movement.

Consistent with mission requirements, vehicles should remain uploaded.

FINAL PREPARATIONS

Preparation for the move should include:

■ Preoperative maintenance checks by vehicle drivers.

■ Reloading of all off-loaded equipment.

■ Complete loading of all service elements; for example, mess, supply, and maintenance.

■ A briefing by the convoy commander to the drivers. It should cover such subjects as safety and instructions addressed in the march order, the tactical situation, and the mission. Strip maps may be issued at this time.

The organization of the march column varies according to the tactical situation and the position area to be occupied. The following points should be considered:

■ The launchers should be dispersed throughout the entire column with at least one buffer vehicle between missiles.

■ Vehicles should be arranged in an order that allows speedy, organized occupation of the new position and defense during movement and occupation.

■ Each vehicle should have an assistant driver, who also serves as an air guard.

■ Machine guns should be distributed evenly throughout the column and aimed alternately to the left and right sides of the route of march.

■ Canvas should be removed so personnel can have their individual weapons poised to return fire if necessary. Unit SOP should specify if some personnel are to fire with their weapons on full AUTO and some with them on SEMIAUTO.

■ Key personnel should be dispersed throughout the column.

■ Fire-fighting equipment should be positioned behind the last launcher in the convoy (except fire extinguishers, which should be with each vehicle/trailer).

■ A means to destroy equipment to prevent enemy use must be readily available.

CONVOY CONTROL MEASURES

The *start point* is normally a geographical feature identifiable on the ground and on a map. The first vehicle of a convoy should cross the start point at the specified time.

A *checkpoint* (CP) is normally a geographical feature identifiable on the ground and on a map. It is used in reporting progress along the route of march.

The *release point* is normally a geographical feature identifiable on the ground and on a map. The last vehicle of a convoy should cross the release point at a specified time. Guides from the advance party meet the convoy at the release point to lead the vehicles into the new position area.

Radio transmissions should be avoided during marches to deny the Threat signal intelligence. Battery SOPs should give procedures for the use of radios, messengers, flags, whistle and horn signals, pyrotechnic signals, and hand and arm signals. (Usage should be in accordance with NATO Standardization Agreement [STANAG] 2154/Quadripartite Standardization Agreement [QSTAG] 539 [as shown in appendix A].) Radio transmissions should be as brief as possible. At no time should specific information that could compromise the

convoy's location be broadcast over the radio; for example, *EXITING THE HIGHWAY* or *PASSING THROUGH TOWN*. References to location should be limited to, for example, *2 KILOMETERS SOUTHWEST OF CHECKPOINT NOVEMBER KILO, HAVE ARRIVED AT RELEASE POINT*.

MARKING THE ROUTE

Road guards may be posted at critical locations where elements of the march might make a wrong turn. However, priority must be given to avoiding detection. Provisions should be made for prompt pickup of the road guards when they are no longer needed. If personnel are not available, route markers can be used. (Be careful not to compromise the unit's location or identity.) Details concerning traffic control and route markings are in FM 19-25 and FM 55-30.

CONDUCT OF THE MARCH

The main body of the heavy or light platoon should move at night, if possible, to avoid location by the enemy. Nuclear convoys must be conducted in accordance with procedures described in chapter 5, in FM 100-50, and by local policies. During movement, the enemy must be prevented from following the convoy to the new position. Speed of the convoy and control of civilian vehicles are necessary to avoid detection. Saboteurs and small, unconventional forces must be prevented from effectively engaging units with small-arms fire.

MARCH DISCIPLINE

Convoys may be organized into serials, which are groupings of march units under separate convoy commanders. The size of the serial should be consistent with the mission and the tactical situation. For example, a serial may consist of one platoon's vehicles, personnel, and equipment; or it may consist of one firing crew's vehicles, personnel, and equipment. The support personnel and equipment of the battery HQ area may deploy as a separate "admin" serial.

Officers and NCOs should ride where they can best control and supervise the march of their units. One of these individuals rides at the head of each serial. The senior person in each vehicle is responsible for ensuring that all orders concerning the march are carried out. All vehicles maintain their order of march unless directed or until circumstances dictate otherwise.

The column should keep moving. The unit SOP should prescribe recovery procedures for disabled vehicles. It should also indicate who stops to pick up mission-essential personnel.

Normally, halts are not scheduled for tactical marches. If a halt is necessary, a wooded area should be selected as the halting place. Vehicles should be dispersed off the road and concealed.

The assistant driver watches for signs, markers, signals, and other traffic. He must also ensure the driver is alert and safely operating the vehicle at all times.

March discipline is attained through training and internal control. The specific objective of march discipline is to ensure cooperation and effective teamwork by march personnel. Teamwork includes:

■ Immediate and effective response to all signals.

■ Prompt relaying of all signals.

■ Obedience to traffic regulations and to the instructions of traffic control personnel.

■ Use of concealment, camouflage, dispersion, radio listening silence, blackout precautions, smoke, and other protective measures against air, ground, and NBC attack.

■ Maintenance of safe speeds, positioning, and intervals between vehicles within the column.

■ Recognizing route marking signals/ signs.

■ Use of correct procedures for handling disabled vehicles.

Radio contact, preferably secure, between lead and trail vehicles should be provided for convoy control.

CONTINGENCIES

IMMEDIATE ACTION PROCEDURES

Because of its nuclear capabilities, Pershing II is always a high-priority target for the Threat. A battery can greatly decrease its vulnerability to attack by establishing an SOP for immediate actions. The following should be considered:

■ Enemy situation—the type of attack that can be expected.

■ Organic resources for countering the different types of attack.

■ Nonorganic support available for countering attacks.

■ Type of communications to be used with the immediate actions, such as flags, radio, and arm and hand signals.

■ How best to protect the unit.

■ How best to neutralize the attack.

AIR ATTACK

If there is an air attack, the unit should disperse to both sides of the road. If terrain offers cover and/or concealment, personnel should dismount and return fire with individual weapons.

AMBUSH

There are two types of ambushes: blocked and unblocked. Both must be countered in the same manner—get out of the kill zone, and neutralize the ambushing force with firepower.

If the route is *blocked*, maximum available fire should be placed on the attacking forces immediately. Personnel in the kill zone should dismount immediately and attack as infantry. The part of the battery not in the kill zone also must react immediately. Unit SOP should clearly detail actions to be taken in this situation.

In an *unblocked* ambush, the battery should increase its speed and move through the ambush area. While moving through, it should place the maximum amount of small-arms and automatic weapons fire on the attackers.

If the area is identified during the map reconnaissance as a likely ambush site, the unit should not pass through the area.

If the ambush or any other enemy action is of a magnitude that will cause the column to be broken up, individual elements should proceed on their own to the new position or rally points as designated by the unit SOP.

UNIT MOVEMENT SOP

Establishing and following a realistic movement SOP will ensure personnel are adequately trained to cope with situations that may confront them. As a minimum, the unit SOP should conform to the battalion SOP and cover the following:

■ Approval authority and requirements for displacing units of the battery for all possible tactical considerations.

■ Duties of convoy commanders.

■ Duties of the courier.

■ Duties of assistant drivers.

■ Convoy organization.

■ Weapons and ammunition to be carried by personnel.

■ Protective equipment to be worn by personnel.

■ Preparation of vehicles (detailed instructions regarding canvases, windshields, tailgates, tie-down procedures, and so forth).

■ Counterambush action.

■ Drills in reaction to air or artillery attack.

■ Security measures (security forces, blackout lights, and so forth).

■ Maintenance and recovery of disabled vehicles.

■ Establishment of rally points.

■ Convoy communications.

SECTION V
OCCUPATION OF POSITION

DELIBERATE OCCUPATION

A deliberate occupation is one that has been planned. An advance party precedes the unit and prepares the position. The occupation may be during daylight hours following a daylight preparation, at night after a daylight preparation, or at night following a nighttime preparation. A common error in a deliberate occupation is allowing too much activity during preparation, thereby risking compromise. Only the minimum number of vehicles and personnel should go forward. When the tactical situation allows, a very good method of occupying a new position is to make the advance preparation before dark and the movement at night. Nighttime movement following a nighttime reconnaissance is often necessary, but it can be more time-consuming.

A guide meets the platoon at the release point and leads the platoon to the entrance of the position area. There, vehicle guides are waiting to lead the vehicles to their selected locations.

The platoon sergeant directs implementation of the security and defense plan as personnel become available.

Survey data should be available from external sources or from a hasty survey.

Additional considerations for night occupations are:

■ Light discipline must be practiced. Proper preparation for a night occupation will minimize the need for lights. Vehicle blackout drive and blackout marker lights should be turned off as soon as the ground guide has begun to lead the vehicle into position.

■ Noise discipline is more important, since noise can be heard at much greater distances at night.

■ The time for occupation is increased.

■ Each vehicle guide should know where his vehicle is in the order of march so the unit can move smoothly into position without halting the column.

■ Red-filtered flashlights are used to lead the vehicles.

CAUTION: Each driver must stop his vehicle whenever he cannot see the light from the guide's flashlight.

■ Vehicles should not be allowed to move within the position without a guide.

HASTY OCCUPATION

The hasty occupation differs from the deliberate occupation mainly in the amount of time available for reconnaissance and preparation. It generally results from unforeseen circumstances and highlights the importance of planning ahead and selecting tentative positions and routes to them. Because of the lack of time for prior preparation, the platoon will need more time for occupation. Therefore, there may be some delay in getting the vehicles off the route of march.

SUSTAINING ACTIONS

Once the occupation is completed and the platoon is in QRA posture, sustaining actions begin. They are continual, and their priorities are determined by the battery commander/platoon leader.

These actions include:

- Improve position defense plans.
- Improve camouflage.
- Bury and/or overhead wire lines.
- Harden critical elements.
- Reposition vehicles.
- Perform maintenance.
- Rehearse reaction forces.
- Refuel.
- Conduct training.
- Resupply all classes of supply.
- Prepare to march order.

Care must be taken in the manner of resupply and refueling, for they can reveal the location of the platoon. If possible, these tasks should be done at night.

The advance party should always be prepared to leave at a moment's notice.

Threat forces have significant capabilities against which Pershing II units must be able to defend. Because of the system's firepower capabilities, Pershing II represents a tactical target of highest priority. A major objective of the Threat will be to seek and destroy Pershing's fire support capabilities with air, airborne, and missile forces. Pershing II units that can be located by agents, radio direction finding, long-range patrols, or airborne radar are much more vulnerable to attack. The positioning of Pershing II units well behind the forward line of own troops (FLOT) will not present a substantial defense from the threat of attack. To accomplish its mission, a Pershing II unit must be able to—

- Avoid detection.
- Communicate.
- Disperse.
- Improve/harden positions.
- Move.
- Defend against small-unit ground attack.
- Defend against airborne attack.
- Defend against air attack.
- Operate in an NBC environment.

ESTABLISHMENT OF PRIORITIES

Thoughout his planning, the platoon leader must consider two possible scenarios and establish his priorities accordingly.

The platoon leader instructed to *continue his mission in the position despite hostile fire* might establish the following tasks in the priority indicated:

- Camouflage.
- Harden critical items of equipment.
- Prepare individual foxholes.
- Prepare defensive positions.
- Select alternate positions and displacement routes, establish a signal should movement be unavoidable, and brief key personnel.

The platoon leader instructed to *displace upon receiving fire* has a different list. For example, before receiving incoming fire, the unit would, in priority—

- Camouflage.
- Prepare limited protection for personnel/equipment.
- Reconnoiter/select alternate positions, displacement routes, and march-order signal.
- Prepare alternate positions.
- Prepare defensive positions.
- Improve individual protection.
- Improve equipment protection.

ORGANIZATION OF THE DEFENSE

The primary consideration in organizing the defense is to provide early warning and defense in depth. To aid the discussion that follows, the illustration below depicts the areas of influence and interest for the Pershing battery/platoon and its supporting infantry.

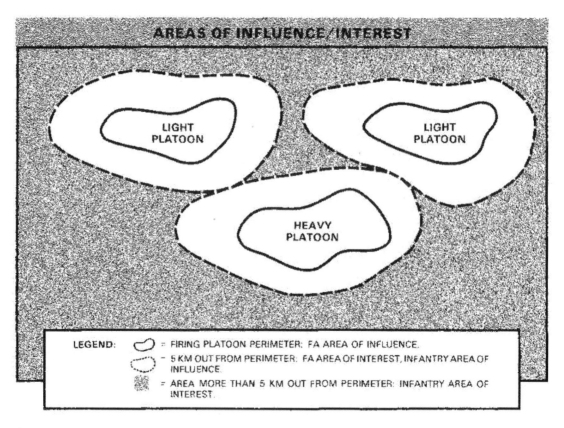

AREAS OF INFLUENCE/INTEREST

LEGEND:
= FIRING PLATOON PERIMETER: FA AREA OF INFLUENCE.
= 5 KM OUT FROM PERIMETER: FA AREA OF INTEREST, INFANTRY AREA OF INFLUENCE.
= AREA MORE THAN 5 KM OUT FROM PERIMETER: INFANTRY AREA OF INTEREST.

COORDINATION

The defense is further organized to include coordination between units, patrols, LPs/OPs, remote sensors, the platoon perimeter, the defense control center, and the reaction force.

The dispersion of Pershing II units on the battlefield allows one unit to provide early warning to another unit. Coordination must ensure rapid transmission of information on Threat activities to all affected units. A spot report using the SALUTE format is an effective means of sending this kind of information.

SALUTE FORMAT	
SIZE:	How many individuals can be seen?
ACTIVITY:	What are they doing?
LOCATION:	Where are they? Give grid coordinates, if possible.
UNIT:	What is their unit? What are the distinctive markings on their uniforms?
TIME:	What time was the sighting made?
EQUIPMENT:	What equipment can you see?

PATROLS

Patrols can effectively provide early warning and engage a small enemy element some distance from the unit's position. These patrols may be mounted or dismounted and should be sent out from the position on an irregular time schedule. (If the unit is in a hide position, mounted patrols will *not* be used.) The DCC must know at all times the positions and routes of patrols so they will not come under fire from the perimeter.

LISTENING/OBSERVATION POSTS

The LPs and OPs should be placed to ensure that a Threat force is identified long before it could threaten the position. Coverage must ensure 360° protection during daylight and darkness. The use of night vision devices will aid in this effort.

Communications must be maintained between LPs/OPs and the defense control center.

REMOTE SENSORS

As time allows, remote sensors are emplaced beyond the limits of the perimeter. Areas likely to be used by the enemy and not easily observable by LPs/OPs or patrols are likely areas for sensor emplacements.

DEFENSIVE PERIMETER

Defensive fighting positions should be chosen to add depth to the defense of the unit's position and to afford intervisibility between fighting positions. The patrols, LPs/OPs, and sensors form the outermost ring of defense. The defensive perimeter must stop the small, squad-size enemy force from successful penetration during the day or night. As time permits, perimeter positions should be dug in. (Guidance on preparing fighting positions is in FM 7-7.) Night vision devices allow the enemy to be seen during the hours of darkness. Although not all perimeter positions need to be continually manned, they must all be designated and improved. The perimeter is divided into quadrants with an NCO-in-charge (NCOIC) assigned for each. One position, usually on a major avenue of approach, within each quadrant is manned at all times.

RANGE CARDS

Each weapon on the perimeter must have a range card. The range card permits the placement of fires on designated targets during periods of limited visibility. It helps in a relief in place by giving the relieving gunner all the information he needs to respond immediately to enemy action. It also gives information to the senior infantryman, first sergeant, and platoon leader for inclusion in their fire planning. Appendix B provides guidance on preparing a range card.

DEFENSE DIAGRAM

After the individual soldiers on the perimeter prepare terrain sketches and range cards, the senior infantryman compiles them into a defense diagram, which he presents to the platoon leader. The firing platoon leader checks the defense diagram for completeness and ensures that the platoon position is adequately defended. Appendix C provides guidance on preparing a defense diagram.

DEFENSE CONTROL CENTER

Within each heavy and light position, a defense control center is established and maintained by the senior infantryman, the firing platoon sergeant, or the battery first sergeant. The DCC is located near the PCC/BCC and serves as the entry control point to the restricted area surrounding the PCC/BCC. A switchboard in the DCC provides communications with the PCC/BCC, the X-gate, listening posts, observation posts, and each perimeter position. All matters concerning the defense

of the position are coordinated through the defense control center.

REACTION FORCE

If the platoon position is attacked or penetrated by enemy forces, the reaction force responds as directed by the defense control center. The reaction force should be made up of at least a squad-type element and should be organized as follows:

■ Reaction force NCOIC—the platoon sergeant or chief of section.

■ One man per firing crew.

■ One man from the PCC/BCC.

■ One man from the communications element.

■ One man from the maintenance element.

■ One man from the mess element.

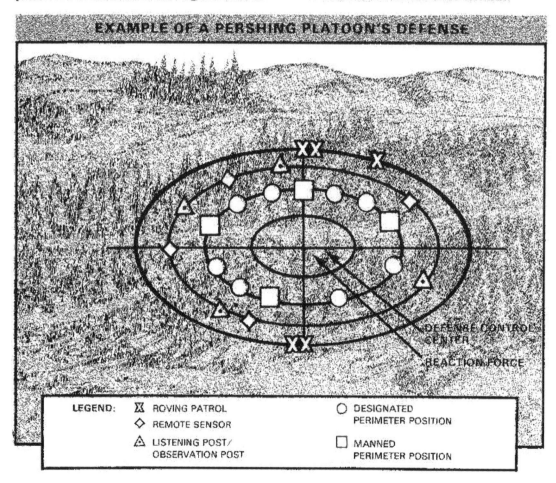

EXAMPLE OF A PERSHING PLATOON'S DEFENSE

LEGEND:
𝕏 ROVING PATROL
◇ REMOTE SENSOR
△ LISTENING POST/ OBSERVATION POST

○ DESIGNATED PERIMETER POSITION
□ MANNED PERIMETER POSITION

DEFENSE AGAINST DETECTION

The most effective means of defending a unit position is to keep the Threat from detecting and locating it. Detection is accomplished through the study of doctrine and signal intelligence, imagery intelligence, and human intelligence. Passive defense measures can greatly enhance survivability.

VISUAL DETECTION

Pershing II units must make every effort to avoid visual detection from the ground and the air.

Night Movement. Whenever possible, vehicles should move at night or during other periods of limited visibility to preclude easy identification of equipment types. Night convoys are harder to follow and identify. An advance party operating during daylight must prevent identification and conceal any Pershing signature.

Equipment Siting. Equipment must be positioned to make maximum use of available natural camouflage. Natural foliage can effectively hide Pershing II units. Positions with deciduous vegetation should

not be selected during the winter. Urban areas may provide excellent concealment and cover.

Camouflage. The camouflaging effect of natural foliage is augmented by erection of lightweight radar-scattering screening systems over equipment. The object is to break the outline of the equipment. Straight, horizontal lines, such as the top of a PCC or a missile, are not natural in a forest and will stand out. The entire piece of equipment must be covered, with particular attention to reflective surfaces such as windshields and mirrors. Individual fighting positions, such as those on the perimeter, are camouflaged much the same way as are vehicles and other pieces of equipment. Camouflage must provide concealment from above as well as from the sides to protect against air and ground observation.

Light and Noise Discipline. Perfect camouflage will not prevent detection unless good light and noise discipline is observed and continuously enforced. Light leaking from a tent or the opened doors of lighted shelters during hours of darkness will expose a unit's position. Noise discipline is very important, especially during the hours of darkness.

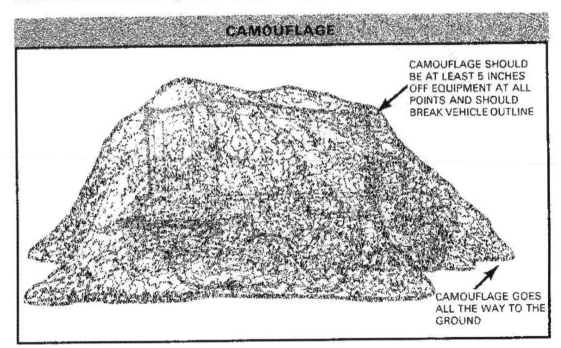

CAMOUFLAGE

CAMOUFLAGE SHOULD BE AT LEAST 5 INCHES OFF EQUIPMENT AT ALL POINTS AND SHOULD BREAK VEHICLE OUTLINE

CAMOUFLAGE GOES ALL THE WAY TO THE GROUND

ELECTRONIC DETECTION

In a nuclear or nonnuclear environment, Pershing II units require multiple communications means with higher headquarters. Radio transmissions must be minimized. Threat forces can locate and identify Pershing II units through the use of radio intercept and radio direction-finding (RDF) techniques. Positioning of units must support defensive communications measures to minimize that possibility.

Positioning of Units. Positioning units parallel, rather than perpendicular, to the FLOT and using directional antennas can help make RDF attempts ineffective. Directing radio signals parallel to the FLOT minimizes those signals that reach Threat forces.

Equipment Siting. Emplacing an antenna so that a hill mass, a group of trees, or a building lies between it and the Threat serves to mask transmissions. The intervening objects either absorb or reflect the radio waves and keep them from reaching the Threat.

Directional Antennas. Directional antennas should be used whenever possible. Transmit only to those people who need to get the information, not to the enemy.

Low Power. Radio sets should be used on the lowest possible output power settings. Since FM radios have a shorter operating range, they should be used rather than AM radios when possible.

Secure Transmissions. All radio operators must be trained to ensure that sensitive or classified information is not transmitted over nonsecure means. Classified information or any essential element of friendly information (EEFI) must be encoded or transmitted by secure means to deny this information to the enemy. The Pershing II EEFI are listed in appendix D.

Short Transmissions. Radio transmissions longer than 10 seconds are highly susceptible to intercept and radio direction finding. A 20- to 25-second transmission is long enough for interception, direction finding of the transmitter, and beginning of the targeting sequence. The targeting sequence can continue even if transmissions cease. Radio operators should be trained to write out messages before keying a microphone and to use frequent breaks in long transmissions.

Wire. If the battery's platoon positions are close enough, wire should be laid between the PCCs and the BCC. If wire line adapters are used, classified information may be transmitted by wire.

Couriers. Couriers should be used to send routine, recurring reports and whenever else possible. Messages may be sent to higher or lower headquarters during routine administrative trips such as those for mess, fuel, and repair parts resupply. Vehicle movement should, as noted before, be during periods of limited visibility and be kept to a minimum. The location of units must be concealed.

DEFENSE AGAINST AIRBORNE ATTACK

Threat airborne forces are targeted specifically against Pershing II. The great mobility of Pershing II units permits movement to avoid engagement with large ground forces if early warning is received. A key defense against airborne forces is to avoid being located. Pershing II units threatened by airborne attack should be moved. Engagement should be undertaken only as a last resort to facilitate withdrawal. Augmentation from rear area forces will be required to neutralize airborne forces.

DEFENSE AGAINST AIR ATTACK

Pershing II units will face, and must be able to defend against, a significant air threat. The best defenses against air attack are to avoid being located by ground-based forces/agents and to avoid being seen from the air. If air defense assets are not available, equipment and personnel should be dispersed to minimize damage from strafing or bombing attacks. If attacked, all organic weapons must be used in accordance with FM 21-2 and FM 44-8. Accuracy is not as important as massed fires. Air defense weapons should be positioned to make best use of their range capabilities.

DEFENSE AGAINST MISSILE/ROCKET ATTACK

Missiles or rockets will be employed against Pershing II units that are accurately located by ground or air forces. Units should be well dug in and trained to respond to nuclear and chemical attacks. Commanders should make every effort to deny information about position locations to the enemy.

DEFENSE AGAINST NUCLEAR, BIOLOGICAL, OR CHEMICAL ATTACK

There is a very real chance that the enemy will use NBC weapons in future conflicts. A detailed discussion of NBC defense is in appendix E and FM 3-5.

REPORTING

Any air or missile/rocket attack must be reported in accordance with STANAG 2008/QSTAG 503 (appendix A) and local directives.

CHAPTER 5
PHYSICAL SECURITY IN NUCLEAR OPERATIONS

Because of the political sensitivity and massive destructive potential of nuclear weapons, physical security for those weapons must be effective. All Pershing II units must be able to provide effective physical security for nuclear weapons in their custody. Physical security of nuclear weapons involves such things as:

■ Selecting only qualified and reliable individuals to work in nuclear operations.

■ Controlling the release (launch) of nuclear weapons.

■ Providing for the security of nuclear weapons in convoy.

■ Safeguarding nuclear weapons at the position area.

■ Destroying equipment to prevent enemy use.

Although every individual in a unit is not assigned as a launcher crewman or custodial agent, all members of the unit must be trained to support the mission of physical security. The most junior member of a unit *could* play a significant role in providing security for a nuclear weapon.

PERSONNEL RELIABILITY PROGRAM

AR 50-5 establishes the personnel reliability program as a safeguard to ensure that only qualified and reliable individuals are allowed to work in nuclear operations.

Personnel will be neither assigned nor trained for assignment to a nuclear duty position until they are screened in accordance with AR 50-5 and found qualified. While in a nuclear duty position, individuals will be screened periodically to ensure that they continue to satisfy the requirements of the program.

It is the reponsibility of the battery commander to ensure that only PRP-qualified individuals are allowed to participate in nuclear operations.

There are two types of nuclear duty positions—controlled and critical. AR 50-5 presents a detailed explanation of each.

In a combat environment, the provisions of AR 50-5 are modified by FM 100-50. For example, the commander may waive administrative procedures of the PRP in combat, including the nuclear duty position roster (NDPR). However, he must take whatever actions are necessary, consistent with good judgment, to fulfill his mission with available manpower resources.

CONTROLS ON NUCLEAR RELEASE

Stringent controls are placed on the handling of nuclear weapons to avoid inadvertent arming and/or launch. These control measures are exercised by the National Command Authority (NCA) and the Joint Chiefs of Staff (JCS). These measures involve the use of sealed authentication systems (SAS) and permissive action links. The policies and procedures for safeguarding and using SAS and PAL are specified in JCS Publication 13.

The permissive action link is a device that interrupts the firing sequence until secure enabling information is received. Procedures must be established to ensure that PAL devices are properly secured in accordance with JCS Publication 13, volume II.

The SAS material is used to authenticate certain nuclear control orders. It must be under two-man control at all times. Procedures involved with SAS material are in JCS Publication 13, volume I.

CUSTODIAL AGENTS

The custodial agents are the direct representatives of the custodian (the last individual who signed for the nuclear weapon). Custodial agents are tasked with maintaining the exclusion area, enforcing the two-man rule, and providing security for the mission vehicle. The training and supervision of custodial agents are command responsibilities from the battery commander down to the first-line supervisor. Custodial agents must be trained in certain areas.

GENERAL KNOWLEDGE

Each custodial agent must have the general knowledge required of a soldier in his grade, such as NBC defense and weapons characteristics. He must thoroughly understand the following concepts.

Two-Man Rule. At least two authorized persons—each capable of detecting incorrect or unauthorized procedures with respect to the task being performed and each familiar with applicable safety and security requirements—must be present during any operation that affords access to material requiring two-man control. Each custodial agent should know when the two-man rule is in effect and the procedures for enforcing it.

Deadly Force. Each custodial agent must know the five circumstances under which the use of deadly force may be authorized. They are as follows:

■ In self-defense, or in defense of another member of the security force, when bodily injury or death is imminent.

■ When necessary to apprehend unauthorized individuals in the vicinity of nuclear weapons or nuclear components.

■ When necessary to prevent unauthorized access, arson, theft, or sabotage of nuclear weapons or nuclear components.

■ When necessary to prevent the escape of an individual believed to have committed one of the above acts.

■ When authorized by a superior because of the occurrence of one of the above acts.

Fire-Fighting Procedures. Each custodial agent must be able to use the fire-fighting techniques for the PII system. He must know the circumstances that would cause him to abandon his attempts to fight the fire. Information on fighting a fire involving a nuclear weapon is in TB 385-2.

KNOWLEDGE OF THE THREAT

Every custodial agent must know the potential threat to the security of the weapons, to include size, tactics, weapons, and means to counter that threat.

SECURITY ASSETS

Each custodial agent must know what security resources are immediately available to him and those that are on call in case of a security incident. He must know the following:

■ The type and amount of ammunition available.

■ The location of the ammunition.

■ The types of reaction forces and the forces that reinforce them.

● The local security devices (to include intrusion detection alarms, remote sensors, and communications nets as applicable).

DESTRUCTION TO PREVENT ENEMY USE

Each custodial agent must know the procedures for destroying his unit's equipment to prevent its use by the enemy. Custodial agents should be trained to perform destruction without supervision if necessary and in the priority determined by the commander on the basis of STANAG 2113/QSTAG 534 (appendix A).

CONVOY OPERATIONS

Custodial agents are trained in all aspects of custody, to include the responsibilities of the courier officer. If for any reason the courier officer no longer can perform his duties, the custodial agents, as designated in the custodial chain of command, will assume the duties of courier officer. If the convoy halts, the courier officer is responsible for maintaining an exclusion area.

EXCLUSION AREA

Nuclear safety rules must be observed while nuclear weapons are kept in open storage. Whenever a platoon has custody of nuclear weapons, it will establish an exclusion area around those weapons. The primary purpose of the exclusion area is to preclude unauthorized or uncontrolled access to the nuclear weapons. In establishing the exclusion area, every effort will be made to provide a safe and secure environment for the weapons while concealing the nature and purpose of the activity. Immediately upon entering an area, guards will dismount from within the convoy and act as a moving exclusion area to prevent unauthorized access to the weapons. Within the defensive perimeter, the exclusion area may range in size from a single missile on an erector-launcher to a basic load of warheads for an entire battery.

Minimum requirements for establishment of an exclusion area are addressed in FM 100-50. More stringent requirements for exclusion area establishment, such as marking a barrier with specific material, may be addressed through local policy.

A single entrance to the exclusion area will be designated. This entrance will be controlled by two PRP-qualified guards. They will allow entry only to authorized personnel as follows:

■ Authorized personnel are those designated on a nuclear duty position roster. A current NDPR will be available to the gate guards.

■ Recognition by the PRP-qualified gate guards is sufficient to allow authorized personnel entry into the exclusion area.

■ The site commander or his designated representative may verbally authorize entry for other personnel.

The number of personnel inside the exclusion area will be accounted for.

Field phone communications from the exclusion area to either the DCC or the BCC/PCC should be maintained at all times.

Within the exclusion area, a limited access area will be established around each weapon system. This limited access area will extend 1 meter out from the erector-launcher in all directions. Entry to this area should be granted only when operations are required in the area. At no time will an individual be alone in the limited access area. All operations in the area will be performed in strict compliance with the two-man rule.

The area outside the exclusion area also must be secured in accordance with chapter 4. This is to preclude unauthorized access to nuclear weapons by preventing unauthorized personnel access to the exclusion area.

LIMITED ACCESS AREA

DUTIES IN THE NUCLEAR CONVOY

Policies and procedures for the logistical movement of nuclear weapons are discussed in AR 50-5. A nuclear convoy transports the weapon from its current location to an alternate storage location or to a firing point. Security and custody of a nuclear weapon are more stringent than for a convoy of conventional ammunition.

CONVOY COMMANDER

Normally, the convoy commander is a commissioned officer who has been tasked to—

■ Ensure security for the convoy is provided.

■ Provide route security.

■ Provide, in coordination with the courier officer, sufficient vehicles and associated equipment for the operation.

■ Ensure drivers and assistant drivers of mission vehicles (those vehicles carrying a nuclear load) have the appropriate clearance for nuclear operations and are in the personnel reliability program.

■ Provide and maintain radio communications with each vehicle in the convoy. In addition, communications will be maintained with elements outside the convoy itself; that is, higher headquarters and additional security forces.

COURIER OFFICERS

The courier officer is the direct representative of the accountable officer or custodian. He has the final authority concerning the security, custody, and destruction to prevent enemy use of the weapon. Normally, he is tasked as follows:

■ Define the exclusion area. The size of the exclusion area for the convoy is at the courier officer's discretion but will never be less than the mission vehicle (to include the cab portion).

■ Inspect all vehicles and equipment involved directly in the transport of the weapon to ensure they are *secure* and *safe*.

■ Provide the material and trained personnel to perform destruction to prevent enemy use during the operation should it become necessary.

■ Inspect all tie-downs, blocking, and bracing for the weapon.

■ Brief all personnel. The courier officer's briefing will include, as a minimum:

□ Custodial chain of command.

□ Purpose (mission) of the convoy.

□ Route of the convoy (to include alternate routes).

□ Emergency actions (destruction, fire fighting, ambush, air attack).

□ Convoy communications procedures.

□ Security requirements.

■ Maintain the equipment necessary to receive, transmit, and authenticate nuclear control orders.

CHECKLIST FOR COURIER OFFICERS

1. BEFORE DEPARTURE

□ Security personnel issued travel orders (if required).

■ Billeting, messing, and return transportation arranged.

□ Security personnel properly equipped.

□ Security personnel checked for appropriate security and reliability clearances.

□ Escort personnel briefed.

□ Strip map with primary and alternate routes provided to drivers. Map includes checkpoints; contact points for military, state, and local authorities; and authorized stops.

□ Administrative documents obtained.

□ COMSEC material obtained.

□ Escort vehicles inspected (as required).

□ Communications check made.

□ Chain of command designated.

□ Appropriate directives on transportation, safety, fire fighting, and security obtained.

□ Security personnel briefed on location of extra ammunition, fire-fighting equipment, and route of march.

■ Guard and transport personnel properly armed and equipped.

■ Transportation equipment inspected by consignee.

□ Security personnel briefed by consignee.

□ Tie-down straps inspected.

□ Materiel loaded and blocked.

□ Materiel signed for.

□ Driver or aircrew issued instructions (DD Form 836 [Special Instructions for Motor Vehicle Drivers] or DD Form 1387-2 [Special Handling Data/ Certification]).

□ Bill of lading and shipping documents obtained.

■ Authorized recipient identified.

□ Convoy formed.

□ Departure time reported.

CHECKLIST FOR COURIER OFFICERS
(CONTINUED)

2. EN ROUTE

☐ Convoy discipline maintained.

☐ One guard, in addition to the driver, present in cab of commercial or military carrier.

☐ Two-man concept in force at all times.

☐ Reports of progress submitted to appropriate headquarters.

3. AT DESTINATION POINT

☐ Arrival reported.

☐ Authorized recipient identified.

☐ Vehicles inspected by authorized recipient.

☐ Equipment unloaded under adequate security.

☐ Receipt for materiel signed by recipient.

☐ Billeting, messing, and return transportation arranged.

Note. *This checklist is provided as a guide only. Other items as directed by higher headquarters or as determined by the courier should be added.*

CONVOY NCOIC

The convoy NCOIC is the courier officer's enlisted assistant. Normally, he is tasked to—

■ Directly supervise the custodial agents.

■ Act as the assistant driver of the mission vehicle or the lead mission vehicle in a multiple vehicle convoy.

■ Act as the courier officer in case of incapacitation or death of the courier officer.

■ Perform duties as assigned by the courier officer.

SECURITY FORCE COMMANDER

Normally, the security force commander is an officer or NCO tasked with training, equipping, and organizing the convoy security forces except the custodial agents. If personnel assets are limited, this position may be filled by the convoy commander.

CONVOY ORGANIZATION

Nuclear convoy organization depends mostly on the tactical situation and the desires of the courier officer. Minimum requirements are in FM 100-50. For planning purposes, the organization must provide for command and control of the convoy, buffer vehicles between mission vehicles, the security force, and a recovery capability.

Minimum personnel requirements for nuclear convoys are specified in AR 50-5 and local policies.

Adequate communications must exist within the convoy. Normally, FM radios are used to control the convoy, and at least one alternate means must be designated for communication with the controlling headquarters. For example, commercial telephone or AM radio could be used as the alternate means.

Test equipment will be carried to verify serviceability of the warhead when it is received from the consignor. This prevents transporting and delivering an unserviceable weapon.

CONVOY SAFETY AND SECURITY

Although the convoy organization depends on the tactical situation, particular emphasis is placed on security and safety. Movement procedures discussed in chapter 3 apply to the nuclear convoy with the following additional considerations:

■ All vehicles must be free of electrical or mechanical defects that could prevent safe arrival. This is determined through proper preventive maintenance checks and services (PMCS).

■ Authorized fire-fighting equipment must be accessible during movement.

■ No maintenance or repairs that might increase the chance of fire will be performed on a vehicle while a nuclear weapon is on board.

■ Excessive handling of nuclear weapons during movement operations will be avoided. Weapons and containers must not be dropped, bumped, or marred.

■ Before movement, all vehicles must be searched and inspected for unauthorized personnel, unauthorized equipment, and sabotage.

■ An exclusion area must be established around the load-carrying vehicle whenever it is stopped, parked, or being loaded or off-loaded.

■ During any movement, communications must be maintained with a headquarters that can respond to a request for assistance.

■ Signals should be developed to ensure personnel are aware of emergency situations as soon as they develop.

AIR MOVEMENT

Nuclear weapons may be moved by air. During air movement, minimum security measures include a courier officer and two PRP-qualified guards for each load-carrying aircraft. The number of different aircrews, couriers, and guards involved in nuclear airlift operations will be kept to the minimum necessary for effective movement. Aircrews must be screened in accordance with AR 50-5.

Times, flight plans, and destinations will be handled on a strict need-to-know basis and will be appropriately classified.

Specific procedures for air movement of nuclear weapons are in FM 100-50. Detailed information on conducting an airlift is in FM 55-12.

DESTRUCTION TO PREVENT ENEMY USE

Destruction to prevent enemy use is authorized in accordance with national and theater policies. The final responsibility for conducting these operations on nuclear weapons, associated documents, and test and handling equipment lies with the highest ranking, on-the-scene individual in the US custodial agent chain of command. Commanders must clearly establish procedures, responsibilities of personnel, training programs, and priorities for destruction.

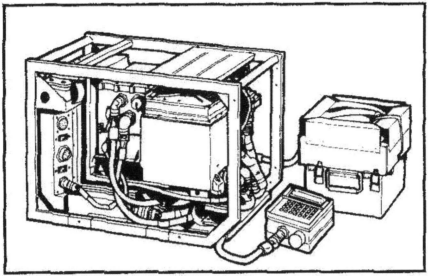

The Pershing II terminal guidance system is functional when a reference scene of the target area is provided. Also, if reference scenes are not available, the Pershing II missile may be fired using only inertial guidance. However, accuracy of the system will be degraded. For the inertial guidance system to be effective, the determined launch location must be within a horizontal accuracy of 50 meters of the actual terrain location. Normally, the individual launch points are determined by the position and azimuth determining system (PADS) team. The battalion has two PADS teams, each consisting of two personnel. The PADS operator, a chief of survey party, and the driver/radio operator have a thorough knowledge of field artillery survey. Each PADS team has a 1/4-ton vehicle to house and transport the PADS equipment. The PADS crews are under operational control of the battalion headquarters. They respond to battery and platoon needs on a mission-by-mission basis.

If PADS is not available, the platoon center must be established by traditional survey techniques. If launch points are not surveyed, proper pacing and computation of pace data must be done to derive each launcher location. All key personnel within the battery should be able to accurately transfer survey control by pacing.

DETERMINATION BY PADS

The PADS may be used in one of two ways to survey firing platoon positions. Each launch point may be located by PADS, or only the platoon center may be located. With either method, each launch point must be located to a horizontal accuracy of 50 meters.

If PADS is used to locate individual launch points, pace data cards are not needed. Before occupation, PADS will locate the launch points designated by the advance party.

If PADS is used to locate the platoon center (PC) only, the location of each launch point must be determined by pacing.

The following limitations should be considered when PADS is used:

■ The PADS must warm up for 30 to 45 minutes before initialization for a survey mission.

■ The PADS must update over a survey control point (SCP) before beginning a survey mission.

■ Each survey must be completed by closing on an SCP within a 55-km radius of the start SCP.

■ Each survey must be completed within 5 hours to avoid inaccuracies.

■ While surveying, PADS must stop at least once every 10 minutes for conduct of a zero velocity correction.

Detailed information on PADS equipment and operations is in FM 6-2 and TM 5-6675-308-12.

Note. The PADS can be mounted in a helicopter to speed the survey process.

SURVEY DATA CARD

Because of operational requirements, the PADS crew may not be available after launch points have been determined by the platoon. Therefore, battalion headquarters operates a survey information center (SIC). The SIC maintains survey information on DA Form 5075-R (Artillery Survey Control Point) for each predesignated position within the battalion area of operation. As necessary, the SIC gives copies of the appropriate DA Form 5075-R to the batteries. Contents of the card give the platoon leader vital information from which launcher locations can be derived. For a description of DA Form 5075-R and its use, see FM 6-2 and STANAG 2865.

COORDINATES

The coordinates used by the PADS team are the complete UTM coordinates that reference the point within the entire grid zone. The easting of the point indicates the number of meters east (or west) the point is from the grid zone's central meridian. The northing of the point is the number of meters that point is north (or south) of the equator (a false northing is used in the southern hemisphere). These coordinates can be found on the map sheet by referencing the small numbers in the lower left-hand margin. However, there is generally no requirement to use the complete coordinates within the firing battery.

For example, the coordinates of the surveyed platoon center for 3d Platoon, Battery B, are 638275.4 5456575.2.

To convert this to missile data, round both the easting and northing to the nearest meter and extract the first five digits to the left of the decimal in both numbers. The grid should read 38275 (easting) 56575 (northing). The platoon leader can then plot this grid as platoon center. For further information on the derivation and structure of the UTM grid system, see FM 21-26.

PACE CARDS

Pacing is used to determine coordinates of launch points when they are not surveyed. A surveyed point (normally the PC) is required for pacing. The surveyed point must be not farther than 500 meters from the launch point for which pacing data will be obtained. When absolutely necessary, pace data may be derived from a prominent terrain feature, such as a road intersection, the grid of which is read from a map. As a class B 1:50,000-scale map is accurate only to within 50 meters horizontally, this is a *last resort*.

CONSTRUCTION OF PACE CARD

Using the DA Form 5075-R provided by the SIC, locate the platoon center or the surveyed point. Using a compass, orient in a cardinal direction (north, south, east, or west) toward the location of the platoon's missiles. Measure the *distance from* the PC along the cardinal direction to a point perpendicular to the launch point of the erector-launcher.

From this position, measure the distance to the base of the round. Measurements may be taken by pacing or by using a steel tape. Your pace *count, or the number of paces per* 100 meters, must be accurately measured.

EXAMPLE:

In the illustration, locate the platoon center and, using *a compass,* orient toward south. Use a factor of 0.9 (111 paces per 100 meters) to convert paces to meters. Pace 25 meters to the south, stop, make a *right face,* and pace 12 meters to the west. This is the location of missile 1. Record the measurements and repeat

the procedure for missiles 2 and 3. Diagram the relative positions of the missiles on a piece of paper and add or subtract the measurements to determine the easting and northing of each missile. The derived UTM coordinates are used as *manual data entries for* each missile.

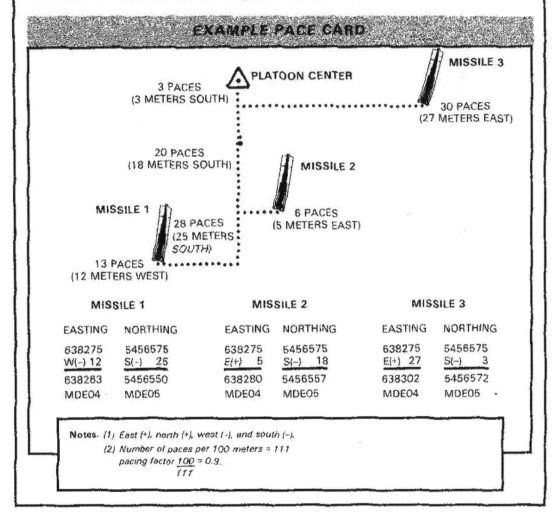

EXAMPLE PACE CARD

MISSILE 1		MISSILE 2		MISSILE 3	
EASTING	NORTHING	EASTING	NORTHING	EASTING	NORTHING
638275	5456575	638275	5456575	638275	5456575
W(-) 12	S(-) 25	E(+) 5	S(-) 18	E(+) 27	S(-) 3
638263	5456550	638280	5456557	638302	5456572
MDE04	MDE05	MDE04	MDE05	MDE04	MDE05

Notes. *(1) East (+), north (+), west (-), and south (-).*

(2) Number of paces per 100 meters = 111

pacing factor $\frac{100}{111} = 0.9$.

DOUBLE CHECK

To avoid erroneous data being programed into the system, two qualified personnel should make independent computations and both sets of data should be checked.

M2 COMPASS

The M2 compass is the primary instrument for determining pace data. The unmounted magnetic compass is a multipurpose instrument used to obtain angle of site and azimuth readings.

COMPONENTS

Azimuth Scale. The azimuth scale is numbered every 200 mils from 0 to 6400, graduated every 20 mils, and can be read to an accuracy of 10 mils.

Sights. The compass has front and rear leaf sights and a mirror in the cover for sighting and reading angles.

Levels. The compass has a circular level for leveling the instrument before azimuth values are read. A tubular level is used with the elevation scale to measure angles of site.

Angle-of-Site Mechanism. Rotation of the level lever causes the elevation level and elevation scale index to rotate as a unit. The index clamps against the bottom piece to keep the mechanism from moving unless actuated by the level lever. This mechanism is rarely, if ever, used in a Pershing unit.

Magnetic Needle and Lifting Mechanism. The magnetic needle provides a magnetic north direction for orienting purposes. The needle is delicately balanced and jewel mounted on a pivot so it rotates freely. The magnetic needle reading is taken when the bubble is centered in the circular level. The lifting mechanism includes a needle-lifting (locking) pin and a needle-lifting lever. The lower end of the pin engages the lever. The upper end projects slightly above the body of the compass to engage the cover when it is closed. Thus, the needle is automatically lifted from its pivot and held firmly against the glass window.

Azimuth Scale Adjuster Assembly. The azimuth scale adjuster assembly rotates the azimuth scale to introduce the declination constant. Two teeth at the adjuster engage teeth on the underside of the azimuth scale so that turning the adjuster with a screwdriver rotates the azimuth scale approximately 1,800 mils. The scale is read against a fixed index under the rear sight hinge.

M2 COMPASS, TOP VIEW

MEASURING AZIMUTH

> CAUTION: When measuring an azimuth, *be sure* no magnetic materials are near the compass.

To read the azimuth scale by reflection, hold the compass in both hands at eye level, with the rear sight nearest your eyes and with your arms braced against your body. Place the cover at an angle of approximately 45° to the face of the compass so you can see the scale reflection in the mirror. Level the instrument by viewing the circular level in the mirror. Sight on the desired object, and read the azimuth indicated on the reflected azimuth scale by the south-seeking end of the compass needle.

CARE AND HANDLING

The M2 compass will not stand rough handling or abuse. Keep the compass in the carrying case, protected from dust and moisture.

M2 COMPASS, OBSERVER'S VIEW

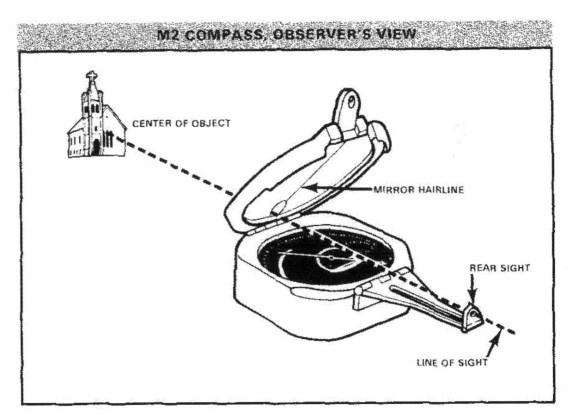

CENTER OF OBJECT

MIRROR HAIRLINE

REAR SIGHT

LINE OF SIGHT

CHAPTER 7
COMBAT SERVICE SUPPORT

Organization, control, and supervision of combat service support (CSS) operations are vital to success of the mission. Providing logistical and administrative support is a significant challenge at the Pershing II firing battery level. Platoons are widely dispersed over large areas. Use of the radio to transmit logistics information must be kept to a minimum. Sustaining the combat effectiveness of each platoon is critical to success. Success depends largely on the ability of the battery, support units, and higher headquarters to provide adequate and timely combat service support. It is achieved through comprehensive planning, coordination, and close supervision of the CSS effort. In CSS operations, every decision must be made to minimize the possibility that units will be seen and located. Unit SOPs must be well thought out, and personnel must be trained on them. Combat service support consists of the logistical and administrative effort required to maintain the battery's capability to fight. The impetus of combat service support is from rear to front or as far forward as the tactical situation will permit. Within the battery, key personnel provide the direction and are responsible for keeping the unit able to fight.

RESPONSIBILITIES

The *battery commander* has overall responsibility for CSS operation and coordinates external support requirements.

The *battery executive officer* establishes priorities for repair and monitors the prescribed load list.

The *battery first sergeant* is responsible for administrative and supply operations.

At the firing platoon level, the *platoon leader* and *platoon sergeant* share the above responsibilities.

CLASSES OF SUPPLY

Knowing the various classes of supply and how to request, procure, store, and distribute supplies is essential to CSS operations.

CLASSES OF SUPPLY

 Class I. Rations and gratuitous issues of health, morale, and welfare items. Included are—
- Meals, ready to eat.
- Expendables such as soap, toilet tissue, and insecticide.
- Water.

 Class II. Clothing equipment, tentage, tool sets and kits, hand tool sets, and administrative and housekeeping supplies/equipment.

 Class III. Petroleum, oils and lubricants.

 Class IV. Construction materials to include camouflage, barrier, and fortification material.

 Class V. Ammunition.

 Class VI. Personal demand items sold through the post exchange.

 Class VII. Major end items, such as trucks.

 Class VIII. Medical supplies.
MAT

 Class IX. Repair parts.

 Class X. Nonstandard items to support nonmilitary programs.

BATTERY TRAINS

The CSS elements of the battery consist of the support platoon, supply section, and mess section. Collectively, these elements may be referred to as the battery combat trains.

In slow-moving situations, the battery trains occupies tactical positions with the heavy platoon. When enemy counterfire is heavy, the battery trains may remain in a central location while the heavy platoon moves from one alternate firing position to another. Wherever the battery trains is located, it must be able to provide responsive support to the firing platoon elements.

The following must be considered in the tactical employment of the battery trains:

■ Cover and concealment.

■ Access to resupply routes and to each firing platoon.

■ Adequate area for resupply and maintenance operations.

■ Location away from main enemy avenues of approach.

■ Defensibility.

■ Alternate positions.

SUPPLY OPERATIONS

Supplies are obtained by unit distribution (battalion/battery delivers) or by supply point distribution (battery/platoon picks up). The battery commander, executive officer, and first sergeant must continually plan for resupply and ensure that it is accomplished. Critical supplies such as rations, POL, ammunition, repair parts, and NBC items must be extensively managed.

RATIONS

From 3 to 5 days supply of combat rations should be kept in the battery. Combat rations are distributed to individuals and crews and carried on organic transportation. When the tactical situation permits, the battery/platoon mess section prepares hot meals. Rations and water are obtained from the battalion through supply point distribution. Rations should be picked up, distributed, and prepared in a way that minimizes the possibility of the unit being seen and identified by enemy agents. Platoons should be able to operate without resupply between unit movements. Extended use of combat rations should be planned. Units should have enough water to preclude the need for resupply between unit moves.

PETROLEUM, OILS AND LUBRICANTS

The POL products include fuels, packaged grease, and lubricants. Each firing platoon has a fuel truck with two 600-gallon tanks, a tank and pump unit that can dispense from either tank, and a trailer-mounted tank. Battery and platoon resupply operations are coordinated with the battalion S4 and carried out in accordance with local policy. Depending on the tactical situation, POL products can be delivered to the battery headquarters area by the maintenance and supply (M&S) company of the maintenance battalion, or a centralized POL resupply point can be established. Within each firing platoon, the vehicles and generators may be topped off in position by the fuel truck or they may be refueled en route to the next position. Each firing platoon should keep enough POL products to preclude need for resupply between unit movements. The SOPs should include safety considerations, basic loads, prevention of contamination, and procedures to take when fuel is contaminated.

AMMUNITION RESUPPLY

Each light and heavy platoon position must be equipped with a basic load of small-arms ammunition for independent operations. Ammunition stockage should ensure that each platoon can defend itself against small ground forces while maintaining a 360° perimeter. Use of smoke may be required for hasty evacuation. The ability to destroy missiles and equipment to prevent enemy use must be continuously ensured. Missile resupply must be done efficiently and must minimize unit vulnerability to being located by enemy agents or being attacked by small-unit ground forces.

REPAIR PARTS

The PLL and essential repair parts stockage list (ERPSL) identify the quantity of combat-essential supplies and repair parts (classes II, IV, and IX) authorized to be on hand or on order at all times. Standardized combat PLLs/ERPSLs are developed on the basis of mandatory parts lists and demands.

The unit PLL clerk manages the PLL and ERPSL and requisitions repair parts when needed.

The PLL/ERPSL loads are maintained so that they are readily accessible to maintenance personnel. Small PLL/ERPSL items are carried in the maintenance truck or trailer. Larger or more bulky items should be carried in vehicles or trailers where the parts may be needed.

The PLL clerk must keep a card file showing the amount and location of all PLL/ERPSL items.

The PLL clerk requisitions repair parts through the battalion maintenance section. From there, requisitions are forwarded to the forward support company.

Repair parts may be delivered to the battery by unit distribution.

OTHER SUPPLIES

Sundry items, such as tobacco, toilet articles, and candy, are issued with rations.

Major end items, such as vehicles, are issued directly to the unit on the basis of tactical priorities and availability of equipment.

Medical supplies are issued through medical channels.

NBC items are class IX in nature but require special emphasis and close management by the battery NBC NCO.

Maps are requisitioned and provided by the battalion S2.

FIELD SERVICES

Field services include laundry, bath, clothing exchange, bakery, textile renovation, salvage, graves registration, decontamination, and post exchange (PX) sales. Conduct of these services must be thoroughly described in local policies and SOPs. Thorough consideration must be given to survivability and operations in an NBC environment.

PERSONNEL SERVICES

Personnel services include personnel management, leaves and passes, postal service, religious activities, legal assistance, financial assistance, casualty and strength reporting, welfare activities, and rest and recreation. Together with ration support and provision of other CSS resources, these services are very important to maintaining morale among the troops. Unit leaders must know how personnel services will be maintained after the outbreak of hostilities. Pershing II unit locations must not be revealed to the enemy during personnel service activities.

HEALTH SERVICES

Health services include health preservation, field sanitation, immunization, medical and casualty evacuation, and safety. To effectively employ the Pershing II system, personnel must be in good physical condition and be able to perform with little rest for extended periods. Field sanitation and safety must be stressed at all levels of command. Immunizations must be kept current. Medical and casualty evacuation must be handled in accordance with local policies. Medical personnel should be in platoon positions, and an ambulance should be at the heavy position. Casualties should be evacuated so as to minimize the possibility of compromising position locations.

MAINTENANCE, REPAIR, AND RECOVERY

Our success on the battlefield is directly related to our ability to keep equipment and materiel in effective operating condition. When breakdowns do occur, equipment must be repaired as far forward as possible by the lowest echelon capable of the repair. When equipment must be moved, it is moved only as far as necessary for repair.

RESPONSIBILITIES

An *operator* must be assigned to each item of equipment. He is responsible for operator-level maintenance using the -10 technical manual.

The *first-line supervisor* supervises the individual operator and crew in maintenance activities.

The *maintenance section* performs battery-level maintenance, which includes minor repairs; performs limited battlefield recovery; and helps in evacuation.

The *motor sergeant* supervises the maintenance section. He ensures necessary repair parts are requisitioned and required test equipment is available. He works directly for the motor officer.

Normally, the *motor officer* is the support platoon leader. He supervises maintenance within the battery and, in conjunction with the XO, establishes priorities for repair.

MAINTENANCE TEAM

The complete maintenance team consists of the operator and/or crew and battery maintenance personnel.

The operator and/or crew performs services on vehicles and equipment as authorized in the -10 technical manual. These include inspecting, servicing, tightening, and lubricating vehicles and equipment and caring for tools. Equipment faults that cannot or should not be repaired by the operator/crew are recorded on a DA Form 2404 (Equipment Inspection and Maintenance Worksheet) and submitted through the first-line supervisor to the battery motor sergeant.

The battery maintenance section performs services listed in the -20 technical manual. These include lubrication services, authorized repairs, road tests, assistance in battlefield recovery, and limited assembly replacement.

Equipment faults not authorized for organizational repair are fixed or replaced by the battalion maintenance section or forward support maintenance company. The forward support company collocated with each FA battalion provides intermediate maintenance support except automotive maintenance. Automotive maintenance support is provided by the maintenance and support company of the maintenance battalion. A forward area support team is attached to each firing battery when in the field to provide on-the-spot maintenance support. Maintenance beyond the capabilities of this team is evacuated to the forward support company location for repair or evacuation to the maintenance battalion.

RECOVERY VEHICLES

The battery has three recovery vehicles. One is normally located with each firing platoon to expedite recovery operations. However, vehicles should be repaired on site, if possible, rather than evacuated. A vehicle stuck in the mud should be recovered by use of its own winch, if possible. FM 20-22 gives detailed information and guidance for all recovery operations.

LOGISTICS RAID SITE

When the tactical situation warrants, a battalion/battery logistics raid site may be established to provide extensive CSS to the battery/platoon. This technique is used when a battery/platoon has been engaged in combat for a sustained period of time, it requires major assistance in several areas at one time, and the supply point distribution method is determined most effective. This technique involves the movement of the battalion/battery CSS elements to a location where the firing elements could pass through the logistics raid site and take on needed supplies, maintenance, and so forth. Following the logistics raid, the CSS elements leave the site.

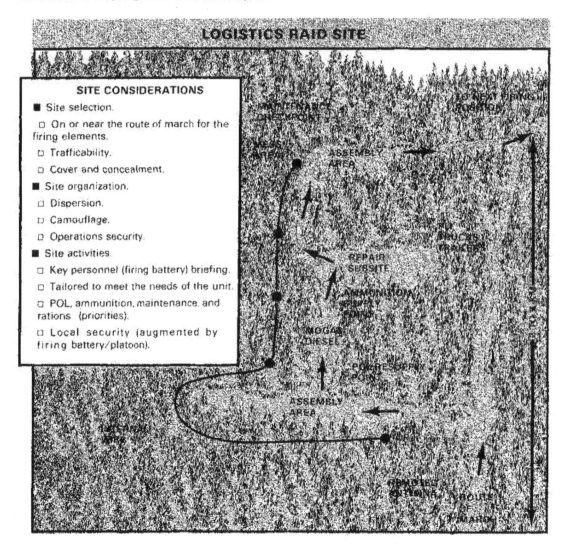

LOGISTICS RAID SITE

SITE CONSIDERATIONS

- Site selection.
  - On or near the route of march for the firing elements.
  - Trafficability.
  - Cover and concealment.
- Site organization.
  - Dispersion.
  - Camouflage.
  - Operations security.
- Site activities.
  - Key personnel (firing battery) briefing.
  - Tailored to meet the needs of the unit.
  - POL, ammunition, maintenance, and rations (priorities).
  - Local security (augmented by firing battery/platoon).

CHAPTER 8
COMMUNICATIONS

The Pershing battery must establish and maintain effective communications both internally and externally. This is a relatively complex task, since firing platoons are located separately from the firing battery headquarters. To accomplish its mission, the Pershing battery must be able to survive on the battlefield and produce timely, accurate fires on designated targets. This requires effective command and control, dissemination of intelligence, and coordination of defense and support operations through a reliable communications system. To ensure responsiveness, reliability, and survivability, all personnel within the battery must understand firing battery communications. Because the operations security (OPSEC) considerations associated with Pershing communications systems are sensitive, a discussion of them is limited to the classified battalion/brigade field manual. This chapter discusses survivability techniques/considerations that apply to any communications system.

COMMUNICATIONS SECURITY

Communications security is designed to deny unauthorized persons information of value that might be derived from a study of our communications. COMSEC includes transmission, cryptographic, and physical security. COMSEC is the responsibility of each individual in the unit. If an individual notes a COMSEC violation, he must report it to his immediate supervisor.

CRYPTOGRAPHIC SECURITY

Cryptographic security is that portion of COMSEC that deals with the proper use of cryptosystems. All classified messages should be transmitted in cryptographic form unless urgency precludes encrypting. Sending a classified message in the clear must be authorized by the commander or his specifically designated representative. The exception to this rule concerns messages classified TOP SECRET. These are *never* sent in the clear over electrical means.

All COMSEC personnel must be carefully trained in the use of the cryptosystems before being permitted to handle actual traffic. AR 380-40 and TB 380-4 should be used to thoroughly familiarize operators of cryptosystems with COMSEC procedures.

Secure communications systems that allow complete freedom and flexibility in the exchange of information are essential to military operations. However, in an emergency, secure communications of any form may not be available and immediate needs may dictate the electrical transmission of classified information in clear text. Information classified TOP SECRET may not be electrically transmitted in the clear over unsecured means at any time. During hostilities, CONFIDENTIAL and SECRET information may be electrically transmitted in the clear by unsecured means (such as telephone, teletypewriter, and radio) as an emergency measure when all of the following conditions exist:

■ The transmitting or receiving station is located in a theater of actual hostilities.

■ Speed of delivery is essential.

■ Encryption is not possible.

■ Transmitted information cannot be acted upon by the enemy in time to influence current operations.

When CONFIDENTIAL or SECRET information must be transmitted in the clear, the following procedures will be followed:

■ Each transmission in the clear must be individually authorized by the commander of the unit or element transmitting the message or by his/her designated representative.

■ References to previously encrypted messages are prohibited.

■ The classification will not be transmitted as part of the message. Messages will be identified by the word "CLEAR" instead of the classification.

■ Each transmission in the clear must be individually authenticated by use of an approved authentication system (transmission authentication).

■ When emergency in-the-clear communications are received, record or other hard-copy messages will be marked "RECEIVED IN THE CLEAR; HANDLE AS CONFIDENTIAL" before delivery to the addressees. In-the-clear messages will be handled as CONFIDENTIAL material and will not be readdressed. Should an addressee determine that the information must be forwarded to another addressee, a new message will be originated, classified, and handled as the subject matter and situation dictate.

VOICE/RADIO TRANSMISSION SECURITY

Radio is the least secure means of transmitting information, yet it is the means most often used in Pershing units. The following paragraphs discuss security procedures that apply to radio transmissions of all types.

RADIO CALL SIGNS

Call signs are changed daily or as often as necessary to deny the enemy information regarding identification and disposition of tactical units (AR 105-64).

ASSIGNMENT AND CHANGES

All call signs have a letter-number-letter configuration and are spoken phonetically (B2E is transmitted *BRAVO TWO ECHO*). These call signs are selected for assignment in a nonpredictable manner. The possibility of duplicated call sign/frequency assignments within an area has been essentially eliminated in the assignment program. The same call sign cannot be used within a corps rear area during its designated period. Higher and adjacent commands and other service elements will be furnished copies of the assigned documents. With the use of nonunique call signs, close liaison is required at all levels. Call signs are assigned to the specific unit and station, not to a net.

Call signs must change at the same time frequencies change.

SUFFIXES

All Army tactical units use changing suffixes (B2E41) along with changing call signs and frequency assignments. The rate of change for suffixes parallels that established for tactical call signs. A simple system has been devised to allow the rapid identification of those "out of net" stations. Instructions have been made for the stations normally operating within a net to use an abbreviated call sign (last letter of letter–number–letter call sign plus suffix number) for most operations. Also, other measures are taken to simplify operating procedures while enhancing security.

ACCIDENTAL COMPROMISE

The purpose of call signs can be defeated if they are associated with other information that may identify the originator or addressee. Compromise can occur when current call signs are used in conjunction with superseded call signs, teletypewriter routing indicators, or corresponding plain-text unit designators appearing in a message.

CLASSIFICATION

Operational call sign assignments of special significance, or those that are changed frequently for security reasons, must be classified at least CONFIDENTIAL.

RADIO FREQUENCIES

Association of radio operating frequencies with a specific unit greatly helps Threat intelligence efforts. The actions cited below are designed to minimize Threat success.

ASSIGNMENT AND CHANGES

Frequencies should be assigned so that specific nets cannot be tracked easily. An identifiable pattern in assignments will be exploited by the Threat on the basis of frequency alone. Operating frequencies should be changed at least once each 24 hours to make continuous intercept difficult. Remember: Call signs must be changed when the operating frequency is changed.

ACCIDENTAL COMPROMISE

If notification of a frequency change must be made by radio, a secure means must be used to prevent accidental compromise.

CLASSIFICATION

Operational frequency assignments of special significance and those that are changed frequently for security reasons must be classified at least CONFIDENTIAL.

OPERATIONS AND PROCEDURES

Adherence to communications procedures as outlined below is critical and will minimize the success of Threat intelligence efforts.

NET DISCIPLINE

Net discipline constitutes operation in accordance with prescribed procedures and security instructions. Such operations include:

■ Correct calling and operating procedures.

■ Use of authorized prosigns and prowords.

■ Operating at a reasonable speed.

■ Correct use of call signs.

■ Effective use of authentication.

■ Limiting transmissions to official traffic.

■ Use of secure devices, when applicable.

AUTHENTICATION

Authentication systems prevent unauthorized stations from entering friendly radio nets to disrupt or confuse operations. They also protect a communications system against false transmissions (imitative deception). The only authentication systems authorized for use in the US Army are those produced by the National Security Agency (NSA) or, for an emergency requirement, by the Intelligence Support Command (INSCOM). If a special or emergency requirement arises, the unit commander must notify the controlling authority of the authentication system in use or the controlling authority's designated representative (communications- electronics [C-E] officer).

The two authorized methods of authentication are challenge-reply and transmission authentication. The operational distinction between the two is that challenge-reply requires two-way communications whereas transmission authentication does not.

Challenge-Reply Authentication. In challenge-reply authentication, the called station always gives the first challenge. This method validates the authenticity of the calling station. It also prevents an unauthorized operator from entering a net to obtain authentication responses for use in another net. The station making the call may counterchallenge the called station, using a different challenge. Only the station responding to a challenge is verified. Do not accept a challenge as an authentication.

If an incorrect reply is received or if an unusual (15- to 20-second) delay occurs between the challenge and the reply, another challenge should be made.

Operators occasionally misauthenticate by using the wrong system or misreading the table. In such cases, the challenging station should try to pinpoint the difficulty and then rechallenge.

The challenge and reply are never given in the same transmission (self-authentication).

Transmission Authentication. This is a method used by a station to authenticate a message. It is used when challenge-reply authentication is not possible; for example, when a station is under radio listening silence. Transmission authentication differs from self-authentication in that authenticators are either carefully controlled to ensure one-time use or are time-based.

PERSONAL SIGNS

Communications operators will not use personal signs, names, and other identifiers while passing traffic.

OFFICIAL MESSAGES

Traffic passed over military communications facilities must be limited to official messages.

EMERGENCY INSTRUCTIONS

Communications operators and supervisors must be familiar with the emergency instructions for their operations. Also, these instructions must be immediately available. Compromised cryptosystems, authentication systems, call signs, and frequency assignments must be superseded as soon as possible. The procedure for supersession must be outlined in unit SOP.

ANTIJAMMING MEASURES

Operators must never reveal over nonsecure circuits an awareness of Threat jamming. Such an admission tells the Threat its effort was effective—an otherwise

---

SITUATIONS REQUIRING CHALLENGE - REPLY OR TRANSMISSION AUTHENTICATION

■ A station suspects imitative deception on any circuit.

■ A station is challenged to authenticate. Stations will not respond to a challenge when under imposed radio listening silence.

■ Radio silence or radio listening silence is directed or a station is required to break an imposed silence (transmission authentication).

■ Contact and amplifying reports are transmitted in plain language.

■ Operating instructions that affect the military situation are transmitted. Examples are closing down a station or watch, changing frequency other than normally scheduled changes, directing establishment of a special communications guard, requesting artillery fire support, and directing relocation of units.

■ A plain language cancellation is transmitted.

■ Initial radio contact is made or contact is resumed after prolonged interruptions.

■ Transmission is to a station under radio listening silence (transmission authentication).

■ A station is authorized to transmit a classified message in the clear.

■ A station is forced, because of no response by the called station, to send a message in the blind (transmission authentication).

extremely difficult determination. Other than physical destruction of Threat equipment, the best antijamming technique is thorough training of communications personnel (FMs 24-18 and 32-20).

ANTENNA SITING

Siting a transmitting antenna on the reverse slope of a hill (away from the Threat) can reduce interception and direction-finding success.

POWER

Minimum power, consistent with operational range requirements, should be used. Power in excess of that required increases vulnerability to interception.

TUNING AND TESTING

For tuning or testing radio communications equipment, a dummy load should be used if available.

NET CONTROL

Adherence to communications procedures is enforced by net control station (NCS) personnel. Transmitted information of potential value to the Threat and deviations from prescribed procedures, which decrease the efficiency of radio nets, can be detected by net control. Depending on the nature of the disclosures, operations can be carried out as planned, revised, or cancelled. In addition, remedial action can be initiated to preclude future occurrences.

DIRECTIONAL ANTENNAS

The use of directional antennas lessens the amount of radio traffic available for Threat forces to intercept.

PHYSICAL SECURITY

Physical security pertains to those measures necessary to safeguard classified communications equipment and material from access by unauthorized persons. Unsuspected physical compromise is far more serious than known loss. If compromise is undisclosed and the cryptosystem continues in use, the Threat may be able to decrypt all traffic sent in that system. Effective physical security ensures the maximum protection of classified material from production to destruction. Classified communications equipment and material can be protected from physical compromise by:

■ Proper handling by all personnel contacted.

■ Adequate storage when not being used.

■ Complete destruction when required.

SECURE AREA FOR COMSEC OPERATIONS

The same physical security principles that apply to a fixed cryptofacility apply to mobile cryptofacilities. While control procedures must be adapted to the field environment, they will be no less stringent than those applied to fixed cryptofacilities. Commanders must ensure the following minimum requirements are met:

■ Guards are the primary means of protection. Cleared operators working inside the cryptofacility fulfill this requirement.

■ If operators are not inside the cryptofacility, the shelter door is secured and guards are provided to patrol the perimeter.

■ A restricted area at least 50 feet in diameter is established and a guard is posted at the entrance to control personnel access.

■ During on-line cryptographic operations, all shelter windows and doors remain closed.

COMSEC CONTROL AND ACCOUNTING

A control system that ensures the proper safeguarding and accounting for classified COMSEC material and information must be established and maintained. The establishment of an accurate and efficient system of accounting is of major importance in preventing the loss or physical compromise of COMSEC information. Such a system should include provisions for registration of the material to be protected, establishment of custodial responsibilities, maintenance of adequate records, and submission of timely and accurate reports. The guidance in TB 380-4 must be followed closely.

DESTRUCTION OF CLASSIFIED WASTE

Work sheets, excess copies, typewriter ribbons, carbon paper, and blotters used in preparing classified information will be handled, stored, and disposed of the same as other classified material.

PACKING AND TRANSPORTATION OF COMSEC MATERIAL

Authorized methods of packing and transporting COMSEC material must meet standards prescribed in AR 380-40. Receipts will be prepared on Standard Form (SF) 153 (COMSEC Material Report) as prescribed by TB 380-4. Transportation of COMSEC material will be by officially designated means.

EMERGENCY EVACUATION OR DESTRUCTION OF COMSEC MATERIAL

To guard against the possible compromise of COMSEC material when a serious threat to the physical security of the material develops, emergency evacuation and destruction plans must be prepared. Various factors, such as the physical features of the area where the material is used, the availability of destruction material, the quantity of material to be destroyed, and the number of personnel available must be considered in the plans.

EMERGENCY EVACUATION PLANS

Plans for emergency evacuation should—

■ Assign specific responsibilities. Assignment should be made by duty position rather than by name, since emergencies can arise at any hour.

■ Ensure that all assigned personnel are familiar with their duties under the plan. Training exercises should be held at regular intervals to ensure that each individual receives detailed instructions on his responsibilities in an emergency.

■ Include information on the location of keys and combinations that might be required in an emergency.

■ Provide for systematic evacuation of COMSEC information to a safe location under the direction of the responsible individual.

Every effort should be made to prevent loss or unauthorized viewing during the period before the return of the information to the original location or its relocation in a new security area.

Factors that would influence the decision to evacuate the COMSEC information include:

■ Time available.

■ Future requirement for the COMSEC information.

■ Degree of hazard involved in the removal.

■ Safety of the new location.

■ Means of transportation available.

■ Transportation routes available.

■ Provisions for precautionary destruction, as outlined in AR 380-40, after the preceding factors are weighed.

EMERGENCY DESTRUCTION PLANS

In the preparation of emergency destruction plans, the factors used in the preparation of emergency evacuation plans will be considered. In addition, the priority of destruction of COMSEC material, as outlined in AR 380-40, will be included.

VIOLATION REPORT

Forward immediately to the controlling authority any report of physical security violations, whether known or suspected, as outlined in AR 380-40 or JCS Pub 13, volume I or volume II, as appropriate.

ELECTRONIC COUNTER-COUNTERMEASURES

In their broad aspects, electronic counter-countermeasures (ECCM) can be considered to deal with means to conceal friendly emitters or to deceive the enemy as to their identity and location. Command posts or weapon systems cannot survive on the battlefield if they are easily identified or located by:

■ The characteristics of their electronic emitters.

■ Incorrect operational procedures.

■ Incorrect practices by users of communications equipment.

Thus, their survival depends on the development and use of good defensive electronic warfare (EW) tactics.

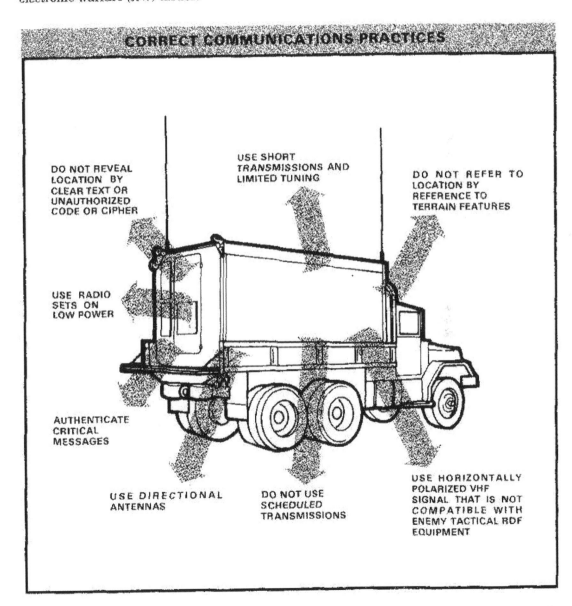

CORRECT COMMUNICATIONS PRACTICES

DO NOT REVEAL LOCATION BY CLEAR TEXT OR UNAUTHORIZED CODE OR CIPHER

USE SHORT TRANSMISSIONS AND LIMITED TUNING

DO NOT REFER TO LOCATION BY REFERENCE TO TERRAIN FEATURES

USE RADIO SETS ON LOW POWER

AUTHENTICATE CRITICAL MESSAGES

USE DIRECTIONAL ANTENNAS

DO NOT USE SCHEDULED TRANSMISSIONS

USE HORIZONTALLY POLARIZED VHF SIGNAL THAT IS NOT COMPATIBLE WITH ENEMY TACTICAL RDF EQUIPMENT

A highly effective way to reduce the chance a signal will be intercepted by the enemy is to reduce communication time. Short communications as a counter-RDF technique are vital during the preparation phase, the approach to a new area of operations, or the defense.

Identifiable electronic signatures and high transmission profiles can result from signal security violations and poor planning before and during an operation. Tactical communications should be used only to—

■ Rapidly convey decisions.

■ Key standing operating procedures.

■ Direct alternative courses of action or changes in mission requirements.

Execution of the unit mission must be inherent in training, planning, and teamwork. Communications are vulnerable to interception and direction finding because of the large volume and the context of communications on command operations nets before a major operation. This pattern uniquely distinguishes a unit in the planning and preparation phase before a movement or a firing. The high communications traffic pattern shown below defies electronic concealment. There is too much traffic to be concealed. To eliminate this indicator, units should strive to achieve a traffic pattern similar to the one depicted in the low traffic pattern, which does not indicate dramatic changes in traffic levels.

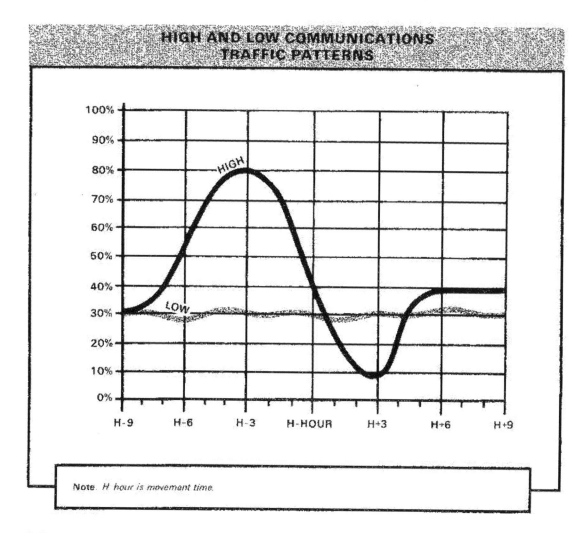

HIGH AND LOW COMMUNICATIONS TRAFFIC PATTERNS

Note. *H hour is movement time.*

New C-E systems with low detectability signatures offer great potential for defeating the Threat radioelectronic combat (REC) capabilities. They make it more difficult to identify and locate critical targets.

The commander has other means available to manage the communications system of his unit:

■ Through the communications-electronics operating instructions (CEOI), specific frequencies and call signs are assigned to specific elements of a command. A frequently changing CEOI is very effective in defeating hostile REC activities. It makes it more difficult for the enemy to identify and exploit the emitters.

■ Emission control restricts use of assigned frequencies to certain critical radio nets or prohibits radio use altogether (partial or complete silence). This tactic keeps the Threat from collecting data on our emissions during a specified period. Also, it greatly reduces the probability of mutual interference between friendly emissions and those of critically important radio nets. Restricting emissions to only a few critical systems may increase the vulnerability of those systems to interception and attack by the enemy. Hence, this tactic should be used with caution.

The use of any radio must be restricted to those individuals who have demonstrated skill in radio communications and are aware of the Threat's REC capabilities. In combat, misuse of the radio through ignorance, lack of training, or carelessness can result in death for an entire unit.

ELECTROMAGNETIC PULSE THREAT

Nuclear strategy and weapons are more sophisticated than ever; and, because of the use of semiconductors and the greater dependence on automatic data processing systems, the significance of the electromagnetic pulse (EMP) threat has increased greatly. It is evident that the EMP effects on tactical communications could be quite severe. Only planned action and proper training to protect against EMP potential effects can preserve the equipment for use when needed. Maintenance and operational practices and procedures to reduce overall vulnerability can be established and practiced. The practical countermeasures listed below will be employed.

ALTERNATE ROUTING

Every supervisor and operator must be trained to establish alternate communications systems if the primary system is disrupted by nuclear effects.

COMMUNICATIONS CABLES

All cables within the command post area should be of minimum length, and there should be no loops that could induce additional energy into the cable. Vans should be relocated if necessary to ensure minimum use of cables. Cables should be buried when possible. When a nuclear strike is imminent, all electrical cables should be disconnected from their associated equipment (generators, radios, and so forth). This will keep EMP from being induced by these conductors and severely damaging equipment.

REPEATING COILS

Repeating coils (C161) must be used on voice and teletype wire circuits.

MAINTENANCE

The integrity of shields, grounding systems, and protective devices must be maintained at all times. Special emphasis should be placed on replacing cables with damaged shielding or connectors.

MESSENGER SERVICE

Either air or motor messenger service should be considered as a primary means of information distribution in an EMP environment until alternative communications can be established.

RADIO ANTENNAS

All antennas and their associated transmission lines (when applicable) should be disconnected from the radio sets when not in use. During periods of listening silence, transmitting antennas should be disconnected. The shields of coaxial cables should be physically grounded to the shelter ground system where the cable enters the shelter.

SHELTER APERTURES

Doors, access panels, and all other apertures should be kept closed as much as possible to reduce the amount of direct EMP in communications shelters.

SPARE EQUIPMENT

Spare equipment should be stored in a protected location.

> **Note.** *Methods of protecting communications equipment from the effects of EMP apply equally to all electronic components, such as missile equipment.*

CHAPTER 9

TRAINING FOR A FIELD ENVIRONMENT

To survive in combat, a unit must train to fight and must continue this training throughout all field exercises and maneuvers. Under the Army training philosophy, the authority and responsibility to organize, conduct, evaluate, and supervise training is delegated through the battalion to the battery level. Decentralized training focuses all training effort at or below the battery level, where the job is actually performed. Therefore, the battery commander is responsible to provide specific instructions to subordinate officers and NCOs to help them prepare and conduct training. Battery officers and NCOs must constantly provide feedback to the battery commander on the battery's training needs and levels of proficiency. See FM 21-6 for guidelines on formulating training.

PERFORMANCE-ORIENTED TRAINING

Performance-oriented training concentrates on those critical tasks that prepare soldiers for combat. Training objectives contain a statement of the task to be performed, the conditions under which the task is performed, and the training standards of acceptable performance. Because of their nature and structure, these objectives facilitate clear and concise thinking about training for combat.

TRAINING PROGRAM DEVELOPMENT

FM 25-2 (Test) is written specifically for training managers. It describes a four-step process to develop training programs. This publication should be used with ARTEP 6-625 and applicable soldier's manuals.

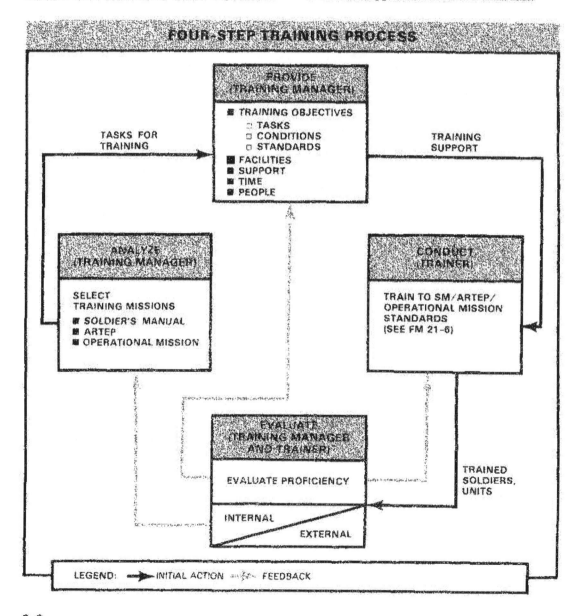

ANALYSIS

In program analysis, the battery commander must determine:

■ The current training level.

■ The desired training level.

■ The means to achieve the desired training level.

The ARTEP and skill qualification test (SQT) standards and results can be used to obtain this information.

PROVISION OF TOOLS

Once the battery commander has completed his program analysis, he determines what tools will best fulfill the training needs.

CONDUCT

The battery commander then programs the training. It will not be possible to select all the performance objectives that were determined in the analysis phase. The BC should focus on those objectives that make the greatest contribution to accomplishment of the unit's mission. He must ensure that qualified personnel conduct the training.

EVALUATION

Performance-oriented training objectives have a built-in evaluation, which shows whether the task has been done to the standards under the given conditions. Thus, the BC can decide whether his unit has reached his training goals. If not, he starts the training cycle again with input from the evaluation. If so, he orients training toward a new goal.

TRAINING MANAGEMENT

Training management is a continuous process. At battery level, the commander must match what he wants to do with what the unit has time and resources to do. The object is to build and sustain readiness of the unit to perform its mission.

TRAINING PUBLICATIONS

The following key training publications form the basis for any battery training program. These publications must be understood and used by all supervisors involved with training or training management:

■ The Army training and evaluation program for the unit (ARTEP 6-625).

■ The soldier's manuals and commander's manuals for all MOSs in the unit.

■ The operator's manual for each piece of equipment in the unit.

■ The applicable technical manuals for the Pershing II system.

■ This field manual.

■ FM 21-6.

■ FM 25-2 (Test).

ARMY TRAINING AND EVALUATION PROGRAM

The ARTEP is designed to assess the unit's ability to—

■ Execute its full wartime mission tasking.

■ Accomplish that mission over a sustained period as a mobile force.

The ARTEP evaluation criteria fall into the following areas:

■ Operations—An assessment of the unit's ability to perform its wartime mission on a sustained basis as a mobile force. This includes attaining and maintaining target coverage, displacing to maintain survivability, and firing simulated missions.

■ Support—An assessment of the unit's capability to use available support resources efficiently in sustained operations. The assessment includes command and control, management, plans, procedures, and proficiency of support personnel. Field maintenance, preventive maintenance, the capability of supply and maintenance personnel to react to equipment breakdowns, and resupply procedures are also assessed.

■ Survival—An assessment of the unit's ability to survive and operate/continue the mission under simulated conditions of enemy attack.

FIELD TRAINING EXERCISE

Successful conduct of a field training exercise (FTX) depends upon a complete scenario based on the ARTEP and good control by battery leaders. Any FTX should be conducted with training to ARTEP standards in mind. The success of the Pershing battery in wartime will depend heavily on the proficiency of individual soldiers, crews, and the battery.

A well-conducted FTX gives the battery commander two well-defined products. First, it demonstrates the ability of the entire unit to perform under simulated combat conditions and identifies requirements for future training. Second, it offers a test of those skills and techniques taught during previous exercises.

SKILL QUALIFICATION TEST

While the ARTEP is critical to the unit, the soldier's manual (SM) and SQT are critical to the individual soldier. A soldier's manual describes what is expected of each soldier at each skill level for that MOS. It contains instructions on how to learn new skills, cites references, and explains the standards that must be met for evaluation (the SQT). The soldier cannot be promoted unless he qualifies for award of the next higher skill level. These individual tasks, when taken collectively, form the ARTEP objectives.

TACTICAL EXERCISE WITHOUT TROOPS

A tactical exercise without troops (TEWT) involves nothing more than a leader taking his subordinates to a predetermined location and discussing application of various tactical principles. During this exercise, the disposition and/or movement of simulated troops or equipment is planned and discussed. Below are some examples of the application of this technique.

EXAMPLE:

One example is a TEWT involving the battery commander accompanied by his XO, first sergeant, platoon leaders, platoon sergeants, and communications chief. The BC would go with these subordinates to different tentative firing positions. An informal, two-way, question-and-answer discussion should be conducted to surface problem areas and resolve them before a tactical exercise. Topics to discuss include:

■ Positioning of erector-launchers.

■ Organization of the position area.

■ Installation of communications.

■ Entrance/exit routes.

■ Locations of LP/OPs.

■ Employment of crew-served weapons.

The BC need not make meticulous preparation for the TEWT; but he must have a firm idea of the tactical principles of reconnaissance, selection, and occupation of a position.

A TEWT may be conducted at the platoon level. In this case, the platoon leader and platoon sergeant go with the assistant chiefs of section (ACS) to a tentative firing position and take the same type of actions as discussed in the previous example.

The TEWT can be a very effective training tool in that it allows the BC/platoon leader to spend a great deal of time with subordinates. Concurrently, subordinate leaders are made aware of problems that could be encountered in a position before an actual tactical exercise.

MAP EXERCISE

A map exercise (MAPEX) is an exercise for unit leaders. It is conducted in a closed environment, such as a classroom, with a simulated wartime scenario and a map of the area of operation. Individual leaders should be assigned the task of orally presenting a detailed, step-by-step plan of what will be done for specific portions of the operation.

The assignments can go out in advance to let the leaders prepare their presentations thoroughly, or the assignments can be given during the exercise to develop the leader's ability to think on his feet.

The exercise may be conducted at battery or battalion level, depending on the desired scope of the exercise.

A MAPEX should be conducted before an FTX, an ARTEP, or other field maneuver and should follow the same scenario; that is, from rollout/load-out through final firing operations in the field.

Through the idea exchange at a MAPEX, many problems may be solved before a unit actually goes to the field.

COMMUNICATIONS EXERCISE

Effective training in Pershing operations depends on effective communications. To provide this support, the communications section must be at a satisfactory training level. A communications exercise (COMMEX) is a means by which this may be sustained. In the COMMEX, the communications elements of the battalion go to the field to establish all necessary communications circuits. These exercises should be conducted often. If the unit's communications assets cannot be sent to the field, training may be done in garrison. This is called a net exercise (NETEX).

NBC TRAINING

To survive in a nuclear or chemical environment, a soldier must be well trained and must be given opportunities to put his NBC skills into practice. One way to maintain NBC proficiency is to integrate NBC situations into field problem scenarios. Another way to train and practice NBC skills is to conduct a round-robin set of NBC training stations. These stations should cover those common-knowledge soldier tasks in the soldier's manual for each MOS in the unit. The soldier should be given a GO or NO GO rating for his performance at each station, with a set standard to be considered NBC proficient. (At least 80 percent on all tasks by 80 percent of the unit is recommended to consider the unit NBC proficient.) A sample list of stations follows:

■ Put on and wear an M17-series protective mask.

■ Perform operator's maintenance on an M17-series protective mask.

■ Identify a chemical agent by using the ABC-M8 or M9 detection paper or the M256 detection kit.

■ Decontaminate self.

■ Decontaminate individual equipment.

■ Recognize symptoms of a nerve agent, and demonstrate related first aid measures.

■ Recognize symptoms of a blood agent, and demonstrate related first aid measures.

■ Read and report radiation dosages.

■ Use the M256 chemical detection kit.

APPENDIX A
INTERNATIONAL STANDARDIZATION AGREEMENTS

Standardization agreements (STANAGs and QSTAGs) are international agreements designed to facilitate allied operations. Upon ratification by the United States, these standardization agreements are binding upon the United States Forces (entirely or with exceptions as noted).

STANAG

A STANAG is the record of an agreement among several or all of the member nations of NATO to adopt like or similar military equipment, ammunition, supplies, and stores and operational, logistic, and administrative procedures. A list of STANAGs in use or under development is published in NATO Allied Administrative Publication (AAP) 4.

QSTAG

A QSTAG is an agreement between two or more ABCA countries (United States, United Kingdom, Canada, and Australia) and is similar in scope to a STANAG. A list of QSTAGs is published in the Quadripartite Standardization Agreement List (QSAL).

APPLICABLE AGREEMENTS

The following STANAGs/QSTAGs are included in this appendix:

■ STANAG 2008/QSTAG 503, Bombing, Shelling, Mortaring and Location Reports.

■ STANAG 2047/QSTAG 183, Emergency Alarms of Hazard or Attack (NBC and Air Attack Only).

■ STANAG 2113/QSTAG 534, Destruction of Military Technical Equipment.

■ STANAG 2154/QSTAG 539, Regulations for Military Motor Vehicle Movement by Road.

STANAG 2008: BOMBING, SHELLING, MORTARING AND LOCATION REPORTS (Edition No. 4) NAVY/ARMY/AIR

ANNEX: A. Format for BOMREP, SHELREP, MORTREP or LOCATION REPORT.

Related Documents: STANAG 2020 – Operational Situation Reports.

STANAG 2103 – Reporting Nuclear Detonations, Biological and Chemical Attacks, and Predicting and Warning of Associated Hazards and Hazard Areas.

AIM

1. The aim of this agreement is to standardize, for the use of the NATO Forces, the method of rendering reports on enemy bombing, shelling, mortaring and locations.

AGREEMENT

2. It is agreed that the NATO Forces are to use the format shown at Annex A when rendering enemy bombing, shelling, mortaring and location reports. (Additional reporting required when NBC weapons are involved is covered in STANAG 2103.) Nations are free to use their own national forms once the basic information has been received by means of the code letters.

3. It is further agreed that the format is to be completed as detailed in the following paragraphs of this agreement.

CLASSIFICATION OF REPORTS

4. Completed reports are to be classified in accordance with current security regulations.

METHOD OF RENDERING AND TRANSMISSION

5. Reports are rendered as normal messages and are to be transmitted by the fastest means available.

CODE WORDS

6. Each transmission is to be preceded by one of the following code words:

   a. SHELREP (in the case of enemy artillery fire).

   b. MORTREP (in case of enemy mortar or rocket fire).

   c. BOMREP (in the case of enemy air attack).

   d. LOCATION REPORT (in the case of location of enemy target). (1)

SECURITY OF MESSAGES

7. The message is always transmitted in clear except as follows:

   a. Unit of Origin—Paragraph A of Annex A. The current call sign, address group or equivalent is to be used.

   b. Position of Observer—Paragraphs B and F.1.b. of Annex A. This is to be encoded if it discloses the location of a headquarters or an important observer post.

   c. When the originator considers that the conditions prevailing warrant a higher classification (e.g., paragraph K, if required).

PARAGRAPHS

8. Each paragraph of the report has a letter and a heading. The headings may be included for each reference to facilitate completion, but only the letters are to be transmitted if the report is sent by radio or telephone.

9. Paragraphs which cannot be completed or are not applicable are omitted from the report.

IMPLEMENTATION OF THE AGREEMENT

10. This STANAG will be considered to have been implemented when the necessary orders/instructions to adopt the method described in this Agreement have been issued to the forces concerned.

Note (1). *To avoid confusion with LOGREP (LOGISTIC REPORT), LOCATION REPORT is written and spoken in full.*

ANNEX A TO STANAG 2008 (Edition No. 4)
FORMAT FOR BOMBING, SHELLING, MORTARING AND LOCATION REPORTS
(SECURITY CLASSIFICATION)
BOMREP, SHELREP, MORTREP OR LOCATION REPORT
(indicate which)

A. **UNIT OF ORIGIN.** Use current call sign, address or group or code name.

B. **POSITION OF OBSERVER.** Grid reference preferred—encode if this discloses the location of a headquarters or important observation posts.

C. **DIRECTION (FLASH, SOUND OR GROOVE) AND ANGLE OF FALL/DESCENT.** (Omit for aircraft.) Grid bearing of flash, sound or groove of shell (state which) in mils, unless otherwise specified. The angle of fall or descent may be determined by placing a stick/rod in the fuze tunnel and measuring in mils, unless otherwise specified, the angle formed by the stick/rod in relation to the horizontal plane.

D. **TIME FROM.**

E. **TIME TO.**

F. **AREA BOMBED, SHELLED OR MORTARED.**

1. Location to be sent as:

  a. grid reference (clear reference is to be used)
      OR
  b. grid bearing to impact points in mils, unless otherwise specified, and distance in meters from observer. This information must be encoded if paragraph B is encoded. (When this method is used, maximum accuracy possible is essential.)

2. Dimensions of the area bombed, shelled or mortared to be given by:

  a. the radius (in meters)
      OR
  b. the length and the width (in meters).

G. **NUMBER AND NATURE OF GUNS, MORTARS, ROCKET LAUNCHERS, AIRCRAFT OR OTHER METHODS OF DELIVERY.**

H. **NATURE OF FIRE.** Adjustment, fire for effect, harassing, etc. (May be omitted for aircraft.)

I. **NUMBER, TYPE AND CALIBER** (State whether measured or assumed.) **OF SHELLS, ROCKETS (OR MISSILES), BOMBS, ETC.**

J. **TIME OF FLASH TO BANG.** (Omit for aircraft.)

K. **DAMAGE.** (Encode if required.)

L. **REMARKS.**

M. **SERIAL NUMBER.** (Each location which is produced by a locating unit is given a serial number.)

N. **TARGET NUMBER.** (If the weapon/activity has previously been given a target number, it will be entered in this column by the locating units.)

O. **POSITION OF TARGET** (the grid reference and grid bearing and distance of the located weapon/activity.)

P. **ACCURACY** (the accuracy to which the weapon/activity located; CEP in meters and the means of location if possible).

Q. **TIME OF LOCATION** (the actual time the location was made).

R. **TARGET DESCRIPTION** (dimensions if possible):

1. radius of target in meters
   OR
2. target length and width in meters.

S. **TIME FIRED** (against hostile target).

T. **FIRED BY.**

U. **NUMBER OF ROUNDS—TYPE OF FUZE AND PROJECTILES.**

*　　　*　　　*　　　*　　　*　　　*　　　*

STANAG 2047: EMERGENCY ALARMS OF HAZARD OR ATTACK (NBC AND AIR ATTACK ONLY) (Edition No. 5) NAVY/ARMY/AIR

ANNEX: A. Alarm Signals

Related Documents:	STANAG 2002 (NBC) -	Warning Signs for the Marking of Contaminated or Dangerous Land Areas, Complete Equipments, Supplies and Stores.
	ATP-45	Reporting Nuclear Detonations, Biological and Chemical Attacks, and Predicting and Warning of Associated Hazards and Hazard Areas.
	STANAG 2104 (NBC) -	Friendly Nuclear Strike Warning.
	STANAG 2889 (ENGR) -	Marking of Hazardous Areas and Routes Through Them.

AIM

1. The aim of this agreement is to provide a standard method of giving emergency alarms within the NATO Forces operating on land, of:

   a. Nuclear, biological or chemical (NBC) hazards and strikes.

   b. Air attack.

AGREEMENT

2. Participating nations agree that NATO Forces, when operating on land, will use the alarm signals detailed herein to give emergency alarms of hazard or attack. Audible and visual alarm signals must be given by means which cannot easily be confused with other sounds or sights encountered in combat. The alarm signals will be given in all cases as soon as an attack or the presence of a hazard is detected. The alarm signals will be repeated throughout the unit area by all who hear or see the original alarm signal, since most available alarm signals are generally limited in range. Additionally, audible and visual alarm signals should normally be supplemented by the simultaneous use of radio, telephone and public address systems.

DETAILS OF THE AGREEMENT

3. It is unlikely that personnel can understand and react quickly and correctly to more than two alarm signals. The following hazards require fast and correct reaction: use or presence of chemical or biological agents, and an imminent air attack or nuclear operation. Therefore, alarm signals for these two hazards are mandatory. (See Note (1).) In the case of radiological contamination, a delay in personnel taking cover may be acceptable.

4. The spoken word (vocal alarm signals) remains the most effective means of informing troops in an emergency.

5. Visual alarm signals are included to supplement the audible alarm signals under conditions when audible signals may be lost due to other noises or to replace audible signals when the tactical situation does not permit the use of sound:

   a. Reliance should not be placed on visual alarm signals during the hours of darkness or in conditions of poor light.

   b. Visual alarm signals should be used when purely audible signals may be lost due to other noise.

   c. Visual signals should be used to warn those personnel arriving at a particular location of an imminent hazard.

   d. Apart from the audio-visual signals detailed at para 2, Note 3 of annex A, normal signal flares are excluded from use as a color alarm signal for NBC and air attack.

6. The actual form of a visual signal and the method of its display are left to the discretion of the local commander. Only the color at annex A is mandatory. However, to aid recognition, whenever possible, the shapes recommended in para 1b and 2b should be used.

7. The alarm signals listed in this agreement are primarily intended to serve as alarms of enemy action. They may be used, however, in an emergency when friendly action could produce similar effects on own forces.

CONFLICT WITH CIVIL REGULATIONS

8. There are some differences between the alarm signals prescribed herein and some national civil defense alarm signals. These differences are considered minor for air attack. Reservations are indicated by each nation where nations or local regulations prohibit NATO forces operating in their territory from sounding alarm signals in exercises and/or alarm signals incompatible with the public warning system in wartime.

PRACTICE ALARM SIGNALS

9. In those cases where nations or local regulations preclude sounding alarm signals during exercises, local commanders should negotiate with local authorities to obtain authorization to sound alarm signals periodically. In the absence of agreement, small alarm devices emitting sound similar to the prescribed audible alarm signals and having limited range should be used during exercises to keep personnel familiar with the audible alarm signals.

Note (1). *No reference is made to ground attack in order to reduce to a minimum the number of signals. Signals for ground attack, if deemed necessary, remain the prerogative of field commanders.*

IMPLEMENTATION OF THE AGREEMENT

11. This STANAG is implemented when the necessary orders/instructions have been issued directing forces concerned to put the content of this agreement into effect.

ANNEX A TO STANAG 2047 (Edition 5)
EMERGENCY ALARMS AND WARNING SIGNS

The following are emergency alarms and warning signs for NATO Forces operating on land:

TYPE OF HAZARD	VISUAL WARNING SIGN	AUDIBLE ALARM SIGNAL
1a. Imminent air attack	1b. Red. Preferably square in shape.	1c. (1) Unbroken warbling siren for one minute.
		(2) Succession of long blasts on vehicle horns, whistles, bugles or other wind instruments in a ratio of 3:1; approximately 3 seconds on and 1 second off.
		(3) Vocal "Air Attack," or corresponding national term where only one nation is involved.
2a. Imminent arrival of or presence of chemical or biological agents or radiological hazards.	2b. (1) Black. Preferably triangular in shape.	2c. (1) Interrupted warbling sound on a siren.
	(2) Donning respirators and taking protective action followed by such hand signals as may be prescribed in local instructions. (See Notes 1, 2, and 3.)	(2) Succession of short signals on vehicle or other horns or by beating metal or other objects in a ratio of 1:1; approximately 1 second on and 1 second off.

TYPE OF HAZARD	VISUAL WARNING SIGN	AUDIBLE ALARM SIGNAL
		(3) Vocal "Gas, gas, gas," or corresponding national term where only one nation is involved.
		(4) Vocal "Fallout, fallout, fallout," or corresponding national term where only one nation is involved.
3a. All clear.	3b. Removal of appropriate warning sign.	3c. (1) Vocal "all clear (specify type of attack)" or corresponding national term when only one nation is involved.
		(2) Steady siren note for one minute or sustained blast on a vehicle horn, whistle, bugle or other wind instrument to indicate absence of all NBC and air attack hazards.

Notes. 1. *Automatic alarms for the early and rapid detection of biological and chemical agents and radiological hazards may complement the devices referred to previously.*

2. *A special audio-visual pyrotechnic signal producing a whistle sound and a yellow, red, yellow display of lights may be used. The combination of colours should be produced as near simultaneously as possible.*

3. *Wearing respiratory protection in the presence of radiological hazards is not mandatory, but will be decided by the local commander.*

*   *   *   *   *   *   *

STANAG 2113: DESTRUCTION OF MILITARY TECHNICAL EQUIPMENT
(Edition No. 3) NAVY/ARMY/AIR

ANNEX: A. Priorities for Destruction of Parts of Military Technical Equipment.

AIM

1. The aim of this agreement is to standardize procedures governing destruction of military technical equipment by the NATO Forces. This equipment does not include medical equipment necessary for the wounded or sick who cannot be evacuated and risk falling into enemy hands.

AGREEMENT

2. Participating nations agree:

a. That it is essential to destroy, to the maximum degree possible, military technical equipment abandoned on land or in harbor during wartime operations, to prevent its eventual repair and use by the enemy.

b. To follow the principles and priorities set forth in this agreement for the destruction of their own equipment when required.

GENERAL

3. *Detailed Methods.* Detailed methods of destroying individual items of equipment are to be included in the applicable publications, user handbooks and drill manuals.

4. *Means of Destruction.* Nations are to provide the means of destroying their own equipment.

5. *Degree of Damage.*

a. General. Methods of destruction should achieve such damage to equipment and essential spare parts that it would not be possible to restore the equipment to a usable condition in the combat zone either by repair or by cannibalization.

b. Classified equipment. Classified equipment must be destroyed to such a degree as to prevent the enemy from duplicating it or learning its method of operation.

c. Associated classified documents. Any classified documents, notes, instructions, or other written material concerning the operation, maintenance, or use of the equipment, including drawings or parts lists, are to be destroyed in a manner which will render them useless to the enemy.

PRIORITIES AND METHODS OF DESTRUCTION

6. *General.*

a. Priority must always be given to the destruction of classified equipment and associated documents.

b. When lack of time or means prevents complete destruction of equipment, priority should be given to the destruction of essential parts, and the same parts are to be destroyed on all similar equipment.

c. A guide to priorities for the destruction of parts for various groups of equipment is contained in Annex A to this STANAG.

7. *Equipment Installed in Vehicles.* Equipment installed in vehicles should be destroyed in accordance with the priorities for the equipment itself.

8. *Spare Parts.* The same priority for destruction of component parts of a major item must be given to the destruction of similar components in spare parts storage areas.

9. *Cryptographic Equipment and Material.* The detailed procedure for the rapid and effective destruction of all types of cryptographic equipment and material is to be specified in the instructions issued by the appropriate authority for security of communications.

10. *Authorization.* The authority for ordering the destruction of equipment is to be vested in divisional and higher commanders, who may delegate it to subordinate commanders when necessary. Standing orders should cover the destruction of isolated equipments which have to be abandoned on the battlefield.

11. *Reporting.* The reporting of the destruction of equipment is to be done through command channels.

IMPLEMENTATION OF THE AGREEMENT

12. This STANAG is considered to be implemented when the principles and priorities indicated herein have been incorporated in appropriate national documents.

ANNEX A TO STANAG 2113 (Edition No. 3)
DESTRUCTION OF MILITARY EQUIPMENT
PRIORITIES FOR DESTRUCTION OF PARTS OF MILITARY
TECHNICAL EQUIPMENT

EQUIPMENT	PRIORITY	PARTS
1. VEHICLES (INCLUDING TANKS AND ENGINEER EQUIPMENT)	1	Carburetor or/fuel pump/injector distributor/fuel tanks/fuel lines.
	2	Engine block and cooling system.
	3	Tires/tracks and suspensions.
	4	Mechanical or hydraulic systems (where applicable).
	5	Differentials/transfer cases.
	6	Frame.
2. GUNS	1	Breech, breech mechanism and spares.
	2	Recoil mechanism.
	3	Tube.
	4	Sighting and fire control equipment (Priority 1 for Anti-Aircraft guns).
	5	Carriage and tires.
3. SMALL ARMS	1	Breech mechanism.
	2	Barrel.
	3	Sighting equipment (including Infra-Red).
	4	Mounts.
4. OPTICAL EQUIPMENT	1	Optical parts.
	2	Mechanical components.
5. RADIO	1	Transmitter (oscillators and frequency generators) and IFF equipment.
	2	Receiver including IFF equipment.
	3	Remote control units or switchboard (exchanges) and operating terminals.
	4	Power supply and/or generator set.
	5	Antennas.
	6	Tuning heads.
6. RADAR AND OTHER ELECTRONIC EQUIPMENT	1	Frequency determining components, records, operating instructions, which are subject to security regulations, and identification material (Identification Friend or Foe [IFF]).
	2	Antennas and associated components such as radiators, reflectors and optics.
	3	Transmission lines and waveguides.
	4	Transmitter high voltage components.
	5	Control consoles, displays, plotting boards.

EQUIPMENT	PRIORITY	PARTS
	6	Cable systems.
	7	Automatic devices.
	8	Other control panels and generators.
	9	Carriage and tires.
7. GUIDED MISSILE SYSTEMS	1	Battery fire control centers.
	2	Missile guidance equipment (including homing systems).
	3	Launchers including control circuits.
	4	Missiles.
	5	Measuring and test equipment.
	6	Generators and cable systems.
8. AIRCRAFT AND SURVEILLANCE DRONES	1	Identification (IFF) equipment, other classified and electronic equipment, publications and documents pertaining thereto, and other material as defined by the national government concerned.
	2	Installed armament (Use subpriorities for Group 2, Guns; or Group 3, Small Arms, as appropriate).
	3	Engine Assembly (Priorities for destruction of magnetos, carburetors, compressors, turbines and other engine subassemblies to be determined by national governments, depending on type of aircraft involved and time available for destruction).
	4	Airframe/control surfaces/undercarriage (Priorities for destruction of propellers, hub-rotor blades, gear boxes, drive shafts, transmissions, and other subassemblies [not already destroyed in Priority 3] to be determined by national governments, depending on type of aircraft involved and time available for destruction).
	5	Instruments, radios, and electronic equipment (not included in Priority 1).
	6	Electrical, fuel and hydraulic systems.
9. ROCKETS	1	Launcher.
	2	Rocket.
	3	Sights and fire control equipment.

*   *   *   *   *   *   *

STANAG 2154/QSTAG 539: REGULATIONS FOR MILITARY MOTOR VEHICLE MOVEMENT BY ROAD (Edition No. 4) NAVY/ARMY/AIR

ANNEXES: A. Definitions
B. Special Movement

Related Documents:	STANAG 1059 –	National Distinguishing Letters for Use by NATO Armed Forces.
	STANAG 2021 –	Computation of Bridge, Raft and Vehicle Classification.
	STANAG 2024 –	Military Vehicle Lighting.
	STANAG 2025 –	Basic Military Road Traffic Regulations.
	STANAG 2041 –	Operation Orders, Tables and Graphs for Road Movement.
	STANAG 2155 –	Road Movement Documents.
	STANAG 2159 –	Identification of Movement Control and Traffic Control Personnel and Agencies.
	STANAG 2174 –	Military Routes and Route/Road Networks.

AIM

1. The aim of this agreement is to set out the basic regulations applying to military motor movement by road for the use of the NATO Forces.

AGREEMENT

2. Participating nations agree to use the regulations applying to military motor movement by road, defined in the following paragraphs, except where they are contrary to national laws and/or regulations.

GENERAL

3. It is particularly important that movement and transport staffs, who are responsible for international road movements and transport are trained to understand and use the terms and definitions listed in annex A.

COLUMNS

4. A column is a group of vehicles moving under a single commander, over the same route, in the same direction.

5. A large column may be composed of a number of organized elements (formations, units or subunits).

6. Each column and organized element of the column should include:

   a. A commander whose location may vary.

   b. In the first vehicle: a subordinate commander known as the pace setter (AAP-6).

   c. In the last vehicle: a subordinate commander known as the trail officer (see annex A).

7. The pace setter of the first element of a column leads it and regulates its speed. The trail officer of the last element deals with such problems as occur at the tail of the column.

8. In addition, each vehicle is to have a vehicle commander (AAP-6) (who may be the driver) who is to be the leader of the vehicle crew appointed for each mission. He is responsible for crew discipline and the execution of the mission.

IDENTIFICATION OF COLUMNS

9. Each column is to be identified in accordance with the laws or regulations of the country within which movement is taking place by flags and/or lights and, in some cases, by a movement number (see annex A).

10. Each column which has received a movement credit (AAP-6) (see para 14 below) is to be identified by a number known as "the movement number" which is allocated by "the authority authorizing/arranging the movement." (See STANAG 2174.) This number identifies the column during the whole of the movement.

11. The movement number is to be placed on both sides and, if possible, on the front of at least the first vehicle and the last vehicle of each organized element of the column. It is to be permanently legible, from ground level, at a minimum distance of 6 meters in normal daylight and composed of:

a. Two figures indicating the day of the month on which the movement is due to commence.

b. Three or more letters indicating the authority organizing the movement, the first two letters being the national symbols of the column. (See STANAG 1059.)

c. Two figures indicating the serial number (AAP-6) of the movement.

d. One letter to identify the element of the column. (This is optional.) Example: Identification of 03-BEA-08-C will indicate that this is the 3rd of the month, moved by BE authority A, as column No. 8, element C.

12. Additionally, each organized element of a column is to be identified by flags and/or for night movement, by lights, security permitting, as described below:

a. The first vehicle of each element of the column is to display a blue flag and a blue light at night if required by national laws or regulations of the country in which the vehicles are operating.

b. The last vehicle of each element of the column is to display a green flag and a green light at night, if required by national laws or regulations of the country in which the vehicles are operating.

c. The vehicle of the column commander is to display a white and black flag as indicated below, subject to the commander's discretion, in certain circumstances.

d. A vehicle that cannot maintain its position in a column should indicate this condition by displaying a yellow flag.

e.   Flags should be approximately 30 cm (12 in) x 45 cm (18 in) in size.

f.   Flags and lights are to be mounted on the front side of the vehicles.

13.   *Headlights.* In peacetime, all vehicles driving in a column are to use dipped headlights (low beam), even in daylight.

MOVEMENT CREDIT

14.   A movement credit is the allocation granted by the authority (see paragraph 10 above) to one or more vehicles in order to move over a controlled route in a fixed time according to movement instructions (see STANAG 2174), a controlled route being a route the use of which is subject to traffic or movement restrictions (AAP-6).

15.   The movement credit includes the indication of times at which the first and the last vehicle of the column are scheduled to pass:

a.   The entry point. (See annex A.)

b.   The exit point. (See annex A.)

c.   Critical points (See annex A.) and, if possible, at traffic control posts.

SPECIAL REGULATIONS FOR THE EXECUTION OF MOVEMENT

16.   All personnel exercising a command in the column and all drivers must obey the instructions of traffic control and regulating personnel.

17.   When approaching a traffic control or a regulating post indicated by prescribed signs (see STANAG 2025) the column commander or his representative must advance ahead of his column and report to the post commander to:

a.   Give the required information concerning formation/unit, route and destination.

b.   Receive information and possible instructions.

18.   Through this post, he can also arrange for the transmission of his own instructions, or information, to the various elements of his column as they pass the post, where however, they must not stop unless ordered to do so.

HALTS

19.   *Short Halts:*

a.   Short halts made by columns or elements of columns on controlled routes normally are to last 10 minutes and in principle should be taken after every 1 hour and 50 minutes running. Wherever possible all columns following the same route should stop at the same time but movement planning must, where necessary, allow at least a 10-minute column gap or gap between columns to ensure that a following column does not overtake the one in front while it is halted.

b.   However, the characteristics of the road may make it necessary for the halt to take place in one particular part of the route rather than simultaneously at a fixed time. In such cases, the necessary instructions are given in the orders relating to the movement.

20.   *Long Halts.* No standard rules for the observance of long halts are laid down. They must always be specifically plotted on movement graphs in order to avoid possible conflict.

21. Particular attention is to be paid to the following aspects of traffic discipline during halts:

a. When making a halt isolated vehicles, or vehicles forming part of a column, should move off the roads as much as possible.

b. If this practice cannot be observed, the commander of a column which is halted must take all necessary measures to facilitate movement of other road users and avoid accidents or traffic jams. The measures to be taken will vary according to the conditions and width of the road and should include:

(1) Warning at a sufficient distance from the front and rear of the column (guards, warning flags, lights or flares), security permitting.

(2) Organizing and directing a system of one-way traffic along the column.

c. When a halted column resumes movement, it has the right of way while moving back on to the road, unless otherwise prescribed.

OVERTAKING OF COLUMNS

22. By isolated vehicles:

a. An isolated vehicle is authorized to overtake a moving column only when:

(1) Its maximum authorized speed is appreciably higher than the speed at which the column is moving, thus enabling it to overtake each vehicle rapidly.

(2) There is sufficient distance between the vehicles of the column to allow the overtaking vehicle to regain its position in the proper lane after overtaking each vehicle.

(3) The trail officer of the column gives a clear signal that overtaking is possible.

b. In all other cases, an isolated vehicle is to overtake the column only when the latter is halted.

23. By other columns:

a. On a controlled route a column may overtake another column only on the orders of the movements authorities and as arranged by the traffic regulating personnel.

b. On an open route no column may overtake another moving column, except in special cases; e.g., on a one-way road which is wide enough. In these cases, the commander of the column desiring to pass is to contact the commander of the column to be passed prior to attempting to pass.

c. Outside these special cases, the overtaking of a column by another column is authorized only if the former is halted and provided the moving column has the time to overtake the whole of the halted column before the latter is ready to move. In this case, the commander of the column desiring to pass is to contact the commander of the column to be passed prior to attempting to pass. The commander of the halted column after giving his agreement, must facilitate the overtaking.

MOVING BY NIGHT

24. *Normal Conditions.* By night, road movement is to be carried out in accordance with national laws and regulations of the country in which the vehicles are operating.

25. *Emergency Conditions.* See STANAG 2024.

ROAD MOVEMENT OF MOTOR VEHICLES/EQUIPMENT

26. The movement by road of certain outsize or heavy vehicles/equipment is restricted by limitations imposed by the different nations. These will call for the application of special procedures and, where necessary, specialized methods, to effect the movement of such equipment and vehicles whether loaded or not.

27. Annex B outlines, with regard to each nation, the class and gage limits beyond which a road movement becomes a special movement.

IMPLEMENTATION OF THE AGREEMENT

28. This STANAG will be considered to have been implemented when the necessary orders/instructions to use the definitions and regulations contained in this agreement have been issued to the forces concerned.

ANNEX A TO STANAG 2154 (Edition No. 4)
REGULATIONS FOR MILITARY MOTOR VEHICLE MOVEMENT
BY ROAD

Definitions

1. Definitions already included in AAP-6:

   a. Those concerning time and distance factors in motor columns:

Column Gap.	*Road Clearance Time.*	*Traffic Flow.*
Traffic Density.	*Road Space.*	
Pass Time.	*Average Speed.*	

   b. Those concerning formation and dispersal of columns:

Start Point.

Release Point.

2. Definitions used for the purpose of this Agreement only:

Blackout lighting. A condition in which lights are so used that they cannot be spotted by enemy observation but which prevent collisions by showing the position of the vehicle to other road users.

Column length. "Column Length" or "Length of a Column" is the length of roadway occupied by a column in movement including the gaps within the column from the front of the leading vehicle to the rear of the last vehicle.

Column. A group of vehicles moving under a single commander, over the same route, in the same direction.

Critical point. That point on a route where any restriction of traffic flow could cause disruption.

Road movement graph. Used by the staffs in planning, supervising and regulating complicated road movements and for providing a convenient means of recording actual moves of units over a period.

APPENDIX B
RANGE CARD

Normally, a soldier prepares a range card while preparing his defensive position. He should complete the range card, except the data section, before constructing the parapet and digging in. This permits the completion of defensive planning as soon as possible. He completes the data section after the fighting position is constructed and the weapon is set in place.

PREPARATION OF THE RANGE CARD

Orient the card (use anything on which you can write, such as a C-ration box top) so both the primary and secondary (if assigned) sectors can be drawn.

Draw a rough sketch of the terrain to the front of the position. Include prominent natural and man-made features that are likely targets, and center the position at the bottom of the sketch.

Fill in the marginal data to include:

■ Gun number.

■ Unit designation.

■ Date.

■ Magnetic north arrow. (Use a compass to determine magnetic north.)

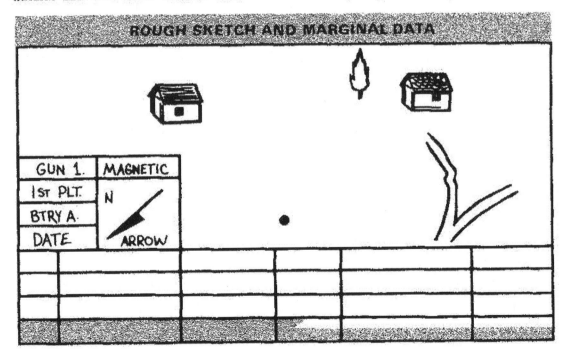

ROUGH SKETCH AND MARGINAL DATA

Specify the location of the gun position in relation to a prominent terrain feature. When no such feature exists, place the eight-digit map coordinates of the position near the point on your sketch representing the position. If there is a prominent terrain feature within 1,000 meters of the gun, use that feature as follows:

■ Using a compass, determine the azimuth in mils from the terrain feature to the gun position.

■ Determine the distance between the gun and the feature. This can be done by pacing or by using a map.

■ Sketch the terrain feature on the card, and identify it.

■ Connect the sketch of the position and the terrain feature with a barbed line from the feature to the gun.

■ Write in the distance in meters (above the line) and the azimuth in mils (below the line) from the feature to the gun.

Sketch in the primary sector of fire with a principal direction of fire (PDF) or a final protective line (FPL) as follows:

■ Primary sector with principal direction of fire:

□ Sketch in the limits of the primary sector of fire as assigned. The sector should not exceed 875 mils (the maximum traverse of the tripod-mounted M60).

□ Sketch in the symbol for an automatic weapon oriented on the most dangerous target within your sector.

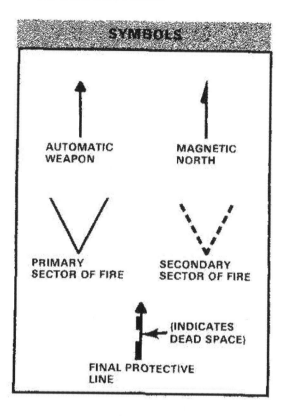

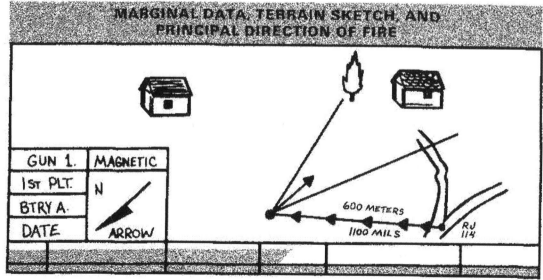

■ Primary sector with final protective line:

□ Sketch in the FPL as assigned. Have someone walk the FPL and determine dead space (sections of the FPL where an individual's waist level is below the line of sight).

□ Reflect dead space on the sketch by a break in the symbol for an FPL. Write in the range to the beginning and end of the dead space.

☐ Write in the maximum range of grazing fire (600 meters for an M60 machine gun if not obstructed at a closer range).

Label the targets in the primary sector in order of priority. Label the FPL or PDF.

Sketch in the secondary sector of fire (as assigned), and label targets within the secondary sector with the range (in meters) from the gun to each.

> **Note.** *The tripod restricts fire to the primary sector. Once emplaced, it should not be moved. The bipod is used when firing into the secondary sector. Sketch in aiming stakes if they are used.*

PREPARATION OF THE DATA SECTION

The tripod and traverse and elevating (T&E) mechanism are used only by the M60 machine gun crew. Use of the T&E mechanism allows the delivery of accurate fires during periods of limited visibility. Information on the T&E mechanism is in FM 7-11B1/2 and FM 23-67. Direction and elevation to a point identified in the sketch as read from the T&E mechanism are entered in the data section of the range card. The range to the point is estimated, read from a map, or measured by pacing and entered in the data section.

USE OF THE RANGE CARD FOR THE M16A1 AND M203

The range card for the M16A1 and M203 is used in conjunction with aiming posts and elbow rests in the fighting position to provide rough alignment of the weapon with the desired target. There is currently no method available for tabulation of target data for these weapons.

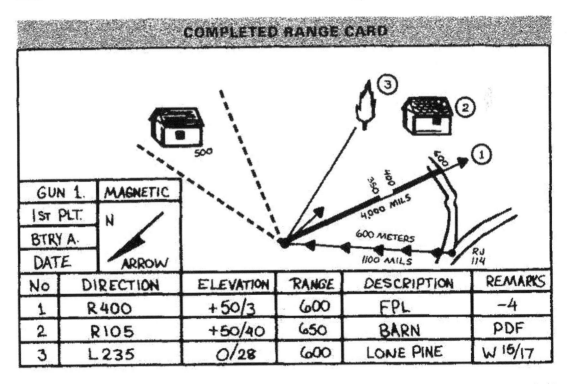

COMPLETED RANGE CARD

GUN 1.	MAGNETIC				
1ST PLT.	N				
BTRY A.					
DATE	ARROW				
No	DIRECTION	ELEVATION	RANGE	DESCRIPTION	REMARKS
1	R 400	+50/3	600	FPL	-4
2	R 105	+50/40	650	BARN	PDF
3	L 235	0/28	600	LONE PINE	W 15/17

The defense diagram is a sketch, drawn to scale, of the platoon defensive resources. It is based on the data from each terrain sketch and range card. It includes the fields of fire for weapons such as M60 machine guns, grenade launchers, and individual weapons. Equipment required for its construction includes a 1:50,000-scale map of the area, a coordinate scale, a protractor, overlay paper, and a blank 1:25,000 grid sheet.

CONSTRUCTION OF THE MATRIX

Locate the platoon center on the 1:50,000-scale map. Identify the grid square or squares that contain the terrain features that influence the defense of the platoon position.

Place tick marks at 200-meter intervals along the sides of the selected grid squares.

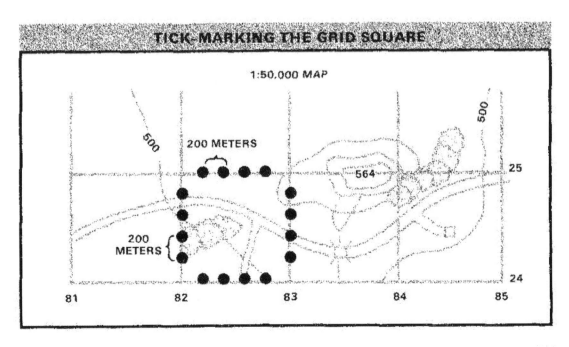

TICK-MARKING THE GRID SQUARE

Connect the tick marks to form 200- by 200-meter squares within each grid square.

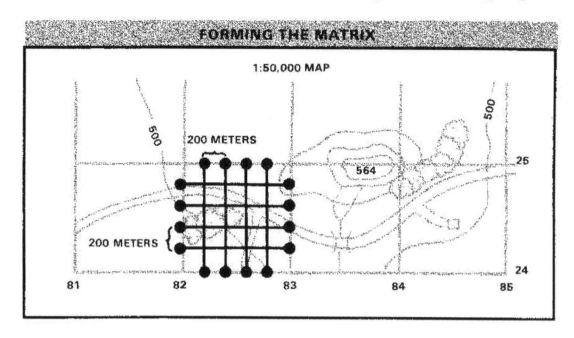

FORMING THE MATRIX

Label the squares beginning at the lower left of the grid square. Number the lines to the right and up, as you would read a map.

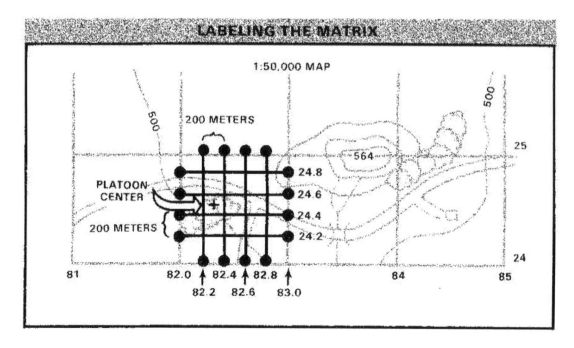

LABELING THE MATRIX

Expand the scale to 1:50,000 by using a blank 1:25,000 grid sheet. In expanding the scale to 1:50,000, each 200- by 200-meter block within the matrix correlates with a 1,000-meter grid square on the 1:25,000 grid sheet. Determine which 200- by 200-meter block on the matrix contains the platoon center. In the figure on the preceding page, it is 82.2 24.4. Next, select a square near the center of the blank grid sheet, and label this square the same—82.2 24.4. You have identified and labeled the 200- by 200-meter block where the platoon center is located. From that point, duplicate the labeling from the matrix on the grid sheet.

Examine the 200- by 200-meter blocks on the 1:50,000 map that contain key terrain features that influence the defense. Sketch what you visualize in these blocks on the corresponding squares of the 1:50,000-scale grid sheet. The result is a map reproduction minus the contour lines and other data not pertinent to the defense of the battery. Contour lines representing hills or depressions may be included if they are deemed pertinent.

PLOTTING ON THE DIAGRAM

Use a 1:50,000 coordinate scale to plot coordinates and to measure distance on the 1:50,000-scale grid sheet by dividing the indicated graduations on the coordinate scale by 10. For example, the 1,000-meter graduation is read as 100 meters. Use the protractor for measuring azimuths or directions.

Plot the weapon locations and sectors of fire. The following sources are available to help in this task:

■ Machine gun range cards indicate the location and the azimuths of the left and right limits and/or final protective line.

■ Individual weapon terrain sketches also may be used to plot areas of coverage.

In all cases, plot locations within 10 meters and directions within 10 mils. On the diagram, indicate the actual location of a weapon by the base of the stem of the weapon symbol. See FM 21-30 for symbols.

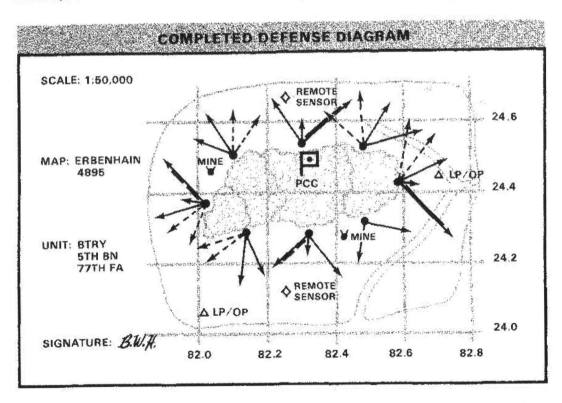

COMPLETED DEFENSE DIAGRAM

SCALE: 1:50,000

MAP: ERBENHAIN 4895

UNIT: BTRY 5TH BN 77TH FA

SIGNATURE: B.W.R.

AVAILABLE RESOURCES

There are many resources available to the battery for its defense. The defense diagram should show where each asset is located. Each platoon leader should check the basic load of ammunition he is authorized and check the battery's field SOP for his share of the battery's assets.

RESOURCES THAT MAY BE AVAILABLE TO A FIRING BATTERY

OBSERVATION DEVICES

Binoculars	20 each
Night vision goggles (AN/PVS-5)	18 each
Night vision sight, individual weapon (AN/PVS-4)	8 each

WEAPONS

Pistol, caliber .45	4 each
Rifle, 5.56-mm (M16A1)	192 each
Grenade launcher, 40-mm (M203)	17 each
Machine gun, 7.62-mm (M60)	16 each
Light antitank weapon (LAW) (M72A2)	30 each
Mine, antipersonnel	18 each
Grenade, fragmentation	108 each
Grenade, smoke	80 each

Essential elements of friendly information are knowledge of a weapon system's operation, capabilities, and vulnerabilities, which, if placed in the wrong hands, could give the Threat a tactical/strategic advantage. Therefore, EEFI must be protected in peacetime as well as during periods of hostility. The distribution statement on the cover of this publication helps protect the information in this manual.

TRAINING

Training of battery personnel in EEFI is the battery commander's responsibility. All personnel are responsible for safeguarding EEFI. Personnel should never discuss unit activities outside their work areas or with people not in their unit. The EEFI should never be discussed over nonsecure means of communication, such as telephones.

PERSHING II EEFI

The following are the EEFI within the Pershing II system:

■ Range of the Pershing II system.

■ Specific performance characteristics that would indicate weaknesses in the system or methods of exploitation.

■ Classified or official use only (FOUO) information related to Pershing.

■ Specific details of planned or employed security measures.

■ Specific details of mobile security measures and SOPs for movement or displacement.

■ Specific plans identifying deployment areas or methods of deployment.

■ Specific electronic security (ELSEC) countermeasures. These include the Pershing II system's signature denial information; for example, masking/screening techniques.

■ Technical countersurveillance measures or plans for countermeasures.

■ The SOPs for field emplacements and/or design peculiarities that cause patterned emplacement in the field.

■ Operating frequencies of radios, ground guidance systems, or radar.

■ Names of key personnel.

■ Extreme personal problems of operators and crewmen.

■ Shortages of personnel.

■ Specific contingency or alert plans.

■ Deployment sequences and times of deployment.

■ Morale problems.

■ Maintenance posture of unit.

■ Equipment shortages.

■ Identification of sensitive equipment or components.

■ Training status.

■ Document identifications (operation orders, plans, contingencies, or any document that would indicate an increased readiness posture).

■ Information regarding communications nets for use in emergencies and/or deployment.

■ Pershing II system vulnerabilities.

■ Communications support unique to reporting.

■ Information on mobilization, locations of assembly areas, control points for emergency usage, and operations centers.

■ Up-to-date classification guide relating to the Pershing II system.

All unit leaders must thoroughly understand the implications of the NBC environment and be prepared to deal with problems as they arise. The NBC defense teams must be well trained to help prepare the unit for operations in an NBC environment. Individuals must be proficient in basic NBC survival skills.

NBC TEAMS

The following special teams, with their inherent functions, will help the battery commander establish a sound unit NBC defense.

BATTERY NBC CONTROL CELL

The NBC control cell, organized under the NBC NCO, advises and helps the commander on prestrike and poststrike NBC defense matters.

Prestrike responsibilities are:

■ Train the survey/monitoring and unit decontamination teams.

■ Train chemical detection teams and operators of chemical agent alarms.

■ Train unit supervisors, who, in turn, train the troops.

■ Ensure that unit NBC equipment is of sufficient quantity, maintained in proper condition, and adequately distributed.

■ Ensure that the unit has a functional, recognizable NBC alarm system in consonance with the applicable STANAG/QSTAG 2047/183 (app A).

■ Advise the battery commander on mission-oriented protective posture.

■ Have available the most current nuclear and chemical downwind messages to predict future NBC hazard areas. Meteorological data is obtained from organic equipment or from higher headquarters.

Poststrike responsibilities are:

■ Supervise survey/monitoring and unit decontamination team operations.

■ Make a simplified fallout prediction.

■ Draw a detailed fallout prediction.

■ Draw a chemical downwind hazard prediction.

■ Send and receive messages and act on NBC 1 through 6 reports.

■ Maintain radiation status of the battery by platoon.

■ Advise the battery commander on where, when, how, and if the unit should move.

PLATOON NBC CONTROL CELLS

Because of their independent locations, each platoon area should have an enlisted alternate to the NBC NCO. This person should be able to do the same tasks as the battery NBC control cell. Normally, he is a member of a light platoon and is trained by the battery NBC NCO.

SPECIAL TEAMS

These teams include the survey/monitoring team for each platoon area and the battery decontamination team. These teams are trained by the NBC officer and NCO. The senior member on each team should act as team chief to immediately supervise the team's operations.

Radiological Survey/Monitoring Teams. The radiological survey/monitoring teams are responsible for nuclear environmental sampling of positions that are occupied, or are to be occupied, by battery units. In each area, there should be a primary team consisting of two trained personnel for each piece of detection or monitoring equipment (IM-93, AN/PDR-27, IM-174). There should be an alternate team for each type of equipment.

Chemical Detection Teams. These teams are responsible for environmental sampling (with M256 kit, M8 alarm, M8/M9 paper) to ensure the unit is not exposed to chemical hazards. The requirements for chemical agent detector operators are the same as those for the radiological detection/monitoring equipment.

Decontamination Teams. The battery also has a decontamination team, made up of at least 10 personnel. They will set up partial decontamination points, help crews in decontamination operations, and help at battalion decontamination points when necessary.

CHEMICAL DEFENSE EQUIPMENT
INDIVIDUAL EQUIPMENT

Each soldier should have at least two complete sets of *chemical protective clothing*, to include the chemical protective overgarment, protective boots, and rubber gloves.

One *protective mask with hood* is issued to each soldier. This, with the chemical protective clothing, affords complete chemical protection. Maintenance of the mask, to include changing of filters, is the responsibility of the individual.

There should be two *M258A1 skin decontamination kits* per individual. The M58A1 training kit can be used to train soldiers on use of the skin decontamination kits.

Other items for individual chemical protection include *M8 detector paper, Mark I injector*, and *M1 canteen cap*. FM 3-4

discusses use of these items. Local commanders will establish issuing policies for chemical agent antidotes.

The *M256 chemical detection kit* is issued to the unit's NCOs, officers, and chemical detection team members. When and where these kits are used are covered in SOPs and the defense plan.

UNIT EQUIPMENT

Normally, there is one M11 decontamination apparatus issued per vehicle and one per crew-served weapon. It is used to spray a liquid decontaminant (DS-2) in partial decontamination. The spray is applied on only those areas with which the operator makes contact during operation. A more thorough decontamination will be performed as soon as time and situation permit.

The chemical/biological (CB) protection system in the PCC and RSGF protects personnel from CB agents. A CB filter cleans CB agents and dust from the air and provides a positive pressure inside the shelter. The M14 CB protective entrance uses a 5-minute air bath to cleanse personnel of vapors. Personnel must be decontaminated before entering the shelter. A control module for the protective entrance is located inside the shelter. The M43 chemical agent detector monitors the air inside the shelter. If contaminants are detected, the M42 chemical agent alarm provides a visible/audible warning signal to personnel inside the shelter.

M8 alarms, consisting of M43 chemical agent detectors, M42 chemical agent alarms, and other ancillary equipment, will be positioned by the unit's NBC team around the position area and on vehicles during convoys. These alarms detect small concentrations of CB agents and warn personnel to assume MOPP level 4. Use of these alarms is covered in FM 3-5 and TM 3-6665-225-12.

NBC OPERATIONS
CHEMICAL ATTACK

Nonpersistent agents dissipate within a few hours. If the agent is present in liquid form, the unit must perform partial decontamination as soon as possible to minimize further spread of the contamination.

Units exposed to persistent agents should be directed by the battalion operations center to move as soon as the mission and tactical situation permit. The unit must immediately go through a decontamination point and undergo complete decontamination of vehicles, equipment, and personnel. Speed and thoroughness are paramount to survivability under these conditions. All personnel and equipment must be decontaminated within 6 hours, which is the limit of effectiveness for the protective clothing. Decontamination operations are discussed in detail in FM 3-5. Personnel and mission-essential equipment should be taken immediately to the decontamination point. Tents, camouflage nets, concertina and communications wire, and so forth, are left in the affected area. If time becomes critical, the unit decontamination team may perform partial equipment decontamination as described in FM 3-5.

As a minimum, when chemical agents are detected in an area, the following actions must take place:

■ Immediately assume MOPP level 4 and sound the alarm (STANAG 2047/QSTAG 183 [app A]).

■ Submit NBC 1 (chemical) report to battalion headquarters by fastest means possible.

■ Use survey/monitoring teams (if not already employed) to detect type, extent, and duration of contamination.

■ Request assistance and permission to move if necessary.

Because of the quick action of modern chemical agents, each individual is responsible for emergency decontamination when he knows or suspects that his skin or personal equipment has become contaminated. Individual decontamination measures are described in FM 3-5.

NUCLEAR ATTACK

If there is a nuclear attack, the unit should immediately implement individual and unit protective measures. Individual perimeter positions, LPs, and OPs should be well dug in. Nuclear fallout may not necessitate a unit movement. If a nuclear blast is observed, the following actions should be taken:

■ Take cover. Deep foxholes with overhead cover provide the best protection.

■ Prepare and send an NBC 1 (nuclear) report in accordance with local policy.

■ Prepare a simplified fallout prediction.

■ Begin continuous radiation monitoring.

■ If radiation is detected, record total dosage received.

■ Remember the following rad tolerances:

□ 70 rads/centigray (cGy)—request, through battalion, permission to move.

□ 150 rads/cGy—radiation sickness can be expected in approximately 5 percent of the unit within 6 hours.

□ 350 rads/cGy or more—there will be deaths within a few days.

□ 650 rads/cGy—100 percent of unit will be combat ineffective within 2 hours; over half will die after approximately 16 days.

■ Decontaminate equipment and personnel as soon as possible.

NBC REPORTING

NBC reporting procedures are discussed in detail in FM 21-40. As a minimum, all senior personnel must be able to prepare and transmit the required reports.

SAMPLE SOP

The following is a portion of a sample SOP for NBC operations. It may be used as a guideline in developing the unit's NBC SOP/field SOP.

SAMPLE BATTERY FIELD SOP
(NBC OPERATIONS PORTION)

APPENDIX 1 (NBC DECONTAMINATION) TO ANNEX F (ACTION TO REDUCE EFFECTS OF ENEMY CHEMICAL AND BIOLOGICAL [CB] ATTACK) TO BATTERY X FIELD SOP

1.   PURPOSE: To outline policy and procedures for personnel and equipment decontamination in Battery X.

2.   DEFINITIONS: There are three types of decontamination.   Each is described below.

   a.   Emergency decontamination is performed by the individual to neutralize and/or remove contaminants from the body.   Emergency decontamination must be performed within 1 minute after contamination for the individual to survive.

   b.   Partial decontamination is performed by unit decontamination teams to remove gross amounts of NBC contamination from weapons, combat vehicles, individual equipment, and other mission-essential equipment. Contaminated personnel remove CB contaminants using personal decontamination kits and portable decontamination apparatuses assigned to vehicles. Radiological contamination is removed by brushing, sweeping, or shaking away dust and debris.   When time is critical, assistance may be provided by the supporting decontamination unit.   Partial decontamination is performed within 1 hour of contamination and outside the contaminated position.

   c.   Complete personnel and equipment decontamination is performed by the chemical detachment with major support from the supported unit. This type of decontamination reduces the NBC contamination hazard to a level that allows soldiers to operate at a lower mission-oriented protective posture.   For chemical decontamination in Battery X, personnel showers will not be given, since the M12A1 power-driven decontamination apparatus (PDDA) will be dedicated to equipment decontamination.   Using fixed undressing procedures, personnel will discard contaminated clothing and equipment and then move to the redressing area for issue of clean clothing.   Personnel with skin contamination will use the M258A1 skin decontamination kit to effect agent removal and decontamination.   Showers will be provided when radiological contamination is involved.

3.   RESPONSIBILITIES:

   a.   Battery and company commanders are responsible for appointing and training a decontamination team and supporting the chemical detachment in performing complete personnel and equipment decontamination.

   b.   The chemical detachment provides support-level decontamination for Battery X and, as necessary, helps units perform partial decontamination.

4.   EXECUTION: Complete personnel and equipment decontamination is a major undertaking and must be carefully planned, coordinated, and executed.

    a.   Predecontamination activities.

        (1) A suitable decontamination site should be jointly selected by the supported unit and the supporting decontamination section. Characteristics of a good decontamination site are: extensive road network trafficable by all equipment and vehicles to be supported, allows water resupply without contact with contaminated equipment, good overhead cover and concealment, adequate water source, capable of being secured, good drainage for runoff, located on a paved road or hardstand and outside the Threat area. Unit commanders will make final determination of location of the decontamination station. Every effort should be made to preselect the site to ensure sufficient time is available to select a site that meets the needs of both the contaminated unit and the supporting chemical detachment. Time is also a critical factor in decontamination operations, since the life cycle of the chemical protective suit is 6 hours in a contaminated environment.

        (2) Identify the chemical agent employed on your position.

        (3) Perform emergency and partial (hasty) decontamination.

        (4) Perform damage control assessment.

        (5) Dispatch medical personnel and unit decontamination team to help set up and operate the decontamination station. At least 12 soldiers are needed to help the chemical detachment. Duties will be: wash with hot soapy water, apply DS-2, apply final rinse, transport personal equipment to redress area, and decontaminate protective masks and other individual equipment.

        (6) Transport squad boxes or laundry bags to decontamination station.

        (7) Emplace contamination markers at contaminated site.

        (8) March order unit according to decontamination priority. Brigade decontamination priorities are:

        (a) Command and control personnel and equipment.

        (b) Missile equipment and crews.

        (c) Missile support equipment and personnel.

        (d) All other.

b.   Actions required at the decontamination site.

(1) On arrival at the decontamination site, the supported unit will maintain command and control of all personnel and equipment to be decontaminated; provide site security (includes security of primary weapon system); perform traffic control; and provide towels, soap, etc.

(2) The chemical detachment will brief key unit personnel on the layout and operation of the decontamination station; call up vehicles, equipment, and personnel for movement through the decontamination station; and direct activities at critical points at the station.

(3) The unit will process through each point of the decontamination station IAW procedures outlined by the chemical detachment. See Tabs A and B.

(4) Key concerns at this point are:

(a) Clean and contaminated areas are clearly marked and observed.

(b) Once decontaminated, neither personnel nor equipment are allowed to reenter the contaminated area of the decontamination station.

(c) No short cuts are taken, since they would endanger the lives of unit personnel and degrade mission capability of the unit.

(d) Provisions have been made for mess, maintenance of equipment, and medical attention.

c.   Postdecontamination activities (the chemical detachment).

(1) Dispose of contaminated clothing and equipment. (When possible, place contaminated clothing and equipment in a pit, cover with layer of super tropical bleach [STB], and bury it.)

(2) Decontaminate detachment personnel and supported unit decontamination team.

(3) Mark area with appropriate contamination markers.

(4) Effect coordination for resupply of DS-2, STB, etc.

(5) Coordinate with controlling battalion for providing personnel showers as time permits.

(6) Continue the mission.

TAB A (PERSONNEL DECONTAMINATION STATION) TO APPENDIX 1 (NBC DECONTAMINA-
TION) TO ANNEX F (ACTIONS TO REDUCE EFFECTS OF ENEMY CHEMICAL AND BIOLOG-
ICAL [CB] ATTACK) TO BATTERY X FIELD SOP

1.   PURPOSE: To outline procedures for processing through a personnel
decontamination station (PDS).

2.   PROCEDURES:

     a.   STEP 1. Contaminated soldiers enter PDS from downwind direc-
tion and decontaminate combat gear (weapon, web gear and harness, etc.)
by submerging it in a solution of STB and water. Once this is completed,
clean soldiers move combat gear to assembly area at the other end of the
PDS.

     b.   STEP 2. Two soldiers move to STB station and wash gross con-
tamination from gloves and overshoes, wash gloves with hot soapy water,
wash mask and hood with hot soapy water.

     c.   STEP 3. Discard overshoes; scrub gloves and boots with hot
soapy water, and rinse.

     d.   STEP 4. Remove combat boots and pass them to members of
supported unit, who transport them to the clean assembly area.

     e.   STEP 5. Remove protective overgarment and place it in desig-
nated container.

     f.   STEP 6. Remove gloves and place them in container.

     g.   STEP 7. Place personal effects (rings, watches, wallets,
etc.) in plastic bags provided by the chemical detachment. Bags are
identified by soldier's dog tags and transported to clean assembly area.

     h.   STEP 8. Remove fatigues/BDUs. (Note: In training exercises,
this step will be conducted in an undressing tent; male and female
soldiers will process separately.) In combat, male and female soldiers
will be processed without regard to gender when gross contamination is
involved.

     i.   STEP 9. Remove underwear and place it in container.

     j.   STEP 10. Take a deep breath, remove mask, rinse it, and pass
it to attendant.

          (1) Supported unit removes filter elements and hood from pro-
tective masks; using a damp sponge, wipes mask with hot soapy water;
using a damp sponge, wipes mask with clear water.

(2) For nuclear contamination, soldiers shower and are monitored for contamination. If contamination is found, soldiers are returned to the shower point and then rechecked.

k. STEP 11. Move to clothing exchange facility.

l. STEP 12. Redress; claim weapon, combat boots, and personal effects; and move to personnel assembly area. Soldiers requiring medical attention move to the medical aid station.

TAB B (EQUIPMENT DECONTAMINATION STATION) TO APPENDIX 1 (NBC DECONTAMINATION) TO ANNEX F (ACTIONS TO REDUCE EFFECTS OF ENEMY CHEMICAL AND BIOLOGICAL [CB] ATTACK) TO BATTERY X FIELD SOP

1. PURPOSE: To outline procedures for processing through an equipment decontamination station.

2. PROCEDURES:

a. STEP 1. Enter site from downwind direction; spray vehicle with hot soapy water. This includes undercarriage and running gears.

b. STEP 2. Move vehicle to DS-2 application station; apply DS-2. Scrubbing may be required to remove thickened chemical agents.

c. STEP 3. Move equipment to holding area for 30-minute contact time.

d. STEP 4. Clean interior of vehicle with hot soapy water.

e. STEP 5. Rinse DS-2 and soapy water from vehicle, and move to agent check station.

f. STEP 6. Check equipment with M256 kit and M8 paper for presence of contamination (use AN/PDR-27 radiac set to check for radiological contamination). If contamination is detected, recycle equipment through decontamination station.

g. STEP 7. Move to maintenance support area to dry and oil equipment. This is essential, since DS-2 is corrosive to metals and softens rubber components.

h. STEP 8. Move to assembly area and link up with personnel.

ABCA—America, Britain, Canada, Australia
AC—alternating current
ACP—allied communications publication
ACS—assistant chief of section
AM—amplitude modulated
ARTEP—Army training and evaluation program

BC—battery commander
BCC—battery control central
BOC—battalion operations center

CB—chemical/biological
C-E—communications-electronics
CEOI—communications-electronics operating instructions
CESO—communications-electronics staff officer
cGy—centigray
COMMEX—communications exercise
COMSEC—communications security
CP—checkpoint
CSS—combat service support

DA Pam—Department of Army pamphlet
DC—direct current
DCC—defense control center
DMA—Defense Mapping Agency

ECCM—electronic counter-countermeasures
EEFI—essential elements of friendly information
EFP—external firing point
EL—erector-launcher
ELSEC—electronic security
EMAS—emergency message authentication system
EMP—electromagnetic pulse
EOL—end of the orienting line
ERPSL—essential repair parts stockage list
EW—electronic warfare

FA—field artillery
FCO—fire control officer

FEBA—forward edge of the battle area
1SG—first sergeant
FLOT—forward line of own troops
FM—frequency modulated, field manual
FOUO—For Official Use Only
FPL—final protective line
FTX—field training exercise

G&C/A—guidance and control/adapter
GIEU—ground integrated electronic unit
GSE—ground support equipment
GTA—graphic training aid

HF—high frequency
HQ—headquarters

ILA—interface logic assembly
INSCOM—Intelligence Support Command

JCS—Joint Chiefs of Staff

km—kilometer

LAW—light antitank weapon
LOGREP—logistic report
LP—listening post

MAPEX—map exercise
MDE—manual data entries
METT-T—mission, enemy, terrain, troops available, and time
MOPP—mission-oriented protective posture
M&S—maintenance and supply
MOS—military occupational specialty
MTOE—modified tables of organization and equipment

NATO—North Atlantic Treaty Organization
NBC—nuclear, biological, chemical
NCA—National Command Authority
NCS—net control station
NCO—noncommissioned officer
NCOIC—NCO in charge

NDPR—nuclear duty position roster
NETEX—net exercise
NSA—national security agency

ODB—operational data base
OIC—officer in charge
OP—observation post
OPSEC—operations security

PAC—Pershing airborne computer
PADS—position and azimuth determining system
PAL—permissive action link
PC—platoon center
PCC—platoon control center
PDDA—power-driven decontamination apparatus
PDF—principal direction of fire
PDS—personnel decontamination station
PID—Pershing identification
PLL—prescribed load list
PMCS—preventive maintenance checks and services
POL—petroleum, oils and lubricants
PRP—personnel reliability program
PII—Pershing II
PX—post exchange

QRA—quick reaction alert
QSAL—Quadripartite Standardization Agreement List
QSTAG—Quadripartite Standardization Agreement

RDF—radio direction finding
REC—radioelectronic combat
RF—radio frequency
RLCU—remote launch control unit
RP—release point
RSGF—reference scene generation facility
RSOP—reconnaissance, selection, and occupation of position
RV—reentry vehicle

SAS—sealed authentication systems
SB—supply bulletin
SC—supply catalog
SCP—survey control point

SF—standard form

SIC—survey information center

SIGSEC—signal security

SM—soldier's manual

SOP—standing operating procedure

SP—start point

SQT—skill qualification test

SSB—single sideband

STANAG—Standardization
Agreement

STB—super tropical bleach

T&E—traverse and elevating
(mechanism)

TACSAT—tactical satellite

TB—technical bulletin

TC—training circular

TDA—tables of distribution and
allowances

TEWT—tactical exercise without
troops

TM—technical manual

TOE—tables of organization and
equipment

UCMJ—Uniform Code of Military
Justice

UTM—universal transverse mercator

X-gate—exclusion area gate

XO—executive officer

REFERENCES

Department of the Army Pamphlets of the 310 series and TM 9–1425–386–L, *List of Applicable Publications (LOAP) for Pershing II Field Artillery Missile System*, should be consulted frequently for latest changes or revisions of references given and for new material on subjects covered in this manual.

REQUIRED PUBLICATIONS

Required publications are sources that users must read in order to understand or to comply with this publication.

ARMY REGULATIONS (AR)

50-5	Nuclear Surety
50-5-1	(C) Nuclear and Chemical Weapons and Material: Nuclear Surety (U)

ARMY TRAINING AND EVALUATION PROGRAM (ARTEP)

6-625	Pershing II Field Artillery Battalion (Projected)

DEPARTMENT OF THE ARMY PAMPHLET (DA PAM)

750-1	Organizational Maintenance Guide for Leaders

FIELD MANUALS (FM)

6-2	Field Artillery Survey
21-2	Soldier's Manual of Common Tasks (Skill Level 1)
21-3	Soldier's Manual of Common Tasks (Skill Levels 2, 3, and 4)
21-40	NBC (Nuclear, Biological and Chemical) Defense
25-2(Test)	How to Manage Training in Units
100-50	Operations for Nuclear-Capable Units

SOLDIER TRAINING PRODUCTS (STP)

6-15E1-SM	Soldier's Manual: Pershing Missile Crewmember (Skill Level 1)
6-15E2/3/4-SM/TG	Soldier's Manual/Trainer's Guide: Pershing Missile Crewmember (Skill Level 2/3/4)
6-21G1/2/3-SM/TG	Soldier's Manual/Trainer's Guide: Pershing Electronics Materiel Specialist (Skill Level 1/2/3)

TABLES OF ORGANIZATION AND EQUIPMENT (TOE)

06625J300	FA Battalion, Pershing II
06627J300	FA Battery, Pershing

TECHNICAL MANUAL (TM)

9-1410-387-12	Operator's and Organizational Maintenance Manual: Pershing II Surface Attack Guided Missile (Projected)

RELATED PUBLICATIONS

Related publications are sources of additional information. They are not required in order to understand this publication.

ALLIED COMMUNICATIONS PUBLICATIONS (ACP)

117	Allied Routing Indicator Book
121	Communications Instructions—General Air/Ground
125(D)	Communications Instructions Radiotelephone Procedures
126	Communications Instructions Teletypewriter (Teleprinter) Procedures
131	Communications Instructions—Operating Signals
134	Telephone Switchboard Operating Procedures

ARMY REGULATIONS (AR)

27-1	Judge Advocate Legal Service
30-1	Army Food Program
40-5	Health and Environment
40-583	Control of Potential Hazards to Health From Microwave and Radio Frequency Radiation
50-101	(C) Safety Rules for the Operation of the Pershing 1A Nuclear Weapon System (U)
55-203	Movement of Nuclear Weapons, Nuclear Components, and Related Classified Nonnuclear Materiel
75-1	Malfunctions Involving Ammunition and Explosives
95-1	Army Aviation: General Provisions and Flight Regulations
105-2	(C) Electronic Counter-Countermeasures (ECCM)--Electronic Warfare Susceptibility and Vulnerability (U)
105-3	Reporting Meaconing, Intrusion, Jamming, and Interference of Electromagnetic Systems
105-10	Communications Economy and Discipline
105-31	Record Communications
105-64	US Army Communications Electronics Operation Instructions (CEOI) Program
165-20	Duties of Chaplains and Commanders' Responsibilities
190-11	Physical Security of Arms, Ammunition, and Explosives
190-28	Use of Force by Personnel Engaged in Law Enforcement and Security Duties
190-40	Serious Incident Report
210-130	Laundry/Dry Cleaning Operations
220-1	Unit Status Reporting
220-58	Organization and Training for Nuclear, Biological, and Chemical Defense
310-1	Publications, Blank Forms, and Printing Management
340-2	Maintenance and Disposition of Records for TOE and Certain Other Units of the Army
350-1	Army Training
350-2	Opposing Force Program
350-4	Qualification and Familiarization With Weapons and Weapons Systems
360-5	Public Information
360-65	Establishment and Conduct of Field Press Censorship in Combat Areas
380-2	High Energy Laser System Technology: Security Classification Guide

380-5	Department of the Army Information Security Program
380-15	(C) Safeguarding Classified NATO Information (U)
380-20	Restricted Areas
380-29	Rocket Propulsion Technology—Security Classification
380-30	(S) Reporting of Critical Intelligence Information (CRITIC) (U)
380-35	(S) Department of the Army Communications Intelligence Security Regulation (Supplement to DOD S-5200.17 [M2] Special Security Manual) (U)
380-40	(C) Policy for Safeguarding and Controlling COMSEC Information (U)
380-150	Access to and Dissemination of Restricted Data
380-200	Armed Forces Censorship
385-10	Army Safety Program
385-64	Ammunition and Explosives Safety Standards
530-1	Operations Security (OPSEC)
530-2	Communications Security
530-3	(C) Electronic Security (U)
530-4	(C) Control of Compromising Emanations (U)
600-10	The Army Casualty System
638-30	Graves Registration Organization and Functions in Support of Major Military Operations
672-5-1	Military Awards
700-65	Nuclear Weapons and Nuclear Weapons Materiel
702-5	Missile Firing Data Reports
703-1	Coal and Petroleum Products Supply and Management Activities
710-2	Supply Policy Below the Wholesale Level
710-9	Guided Missile and Large Rocket Ammunition Issues, Receipts, and Expenditures Report
725-50	Requisitioning, Receipt, and Issue System
735-5	Basic Policies and Procedures for Property Accounting
735-11	Accounting for Lost, Damaged, and Destroyed Property
750-25	Army Test, Measurement, and Diagnostic Equipment (TMDE) Calibration and Repair Support Program
750-40	Missile Materiel Readiness Report

DEPARTMENT OF ARMY (DA) FORMS

| 2404 | Equipment Inspection and Maintenance Worksheet |
| 5075-R | Artillery Survey Control Point |

DEPARTMENT OF ARMY PAMPHLETS (DA PAM)

1-2	Personnel Administrative Center (PAC): Guide for Administrative Procedures
50-3	The Effects of Nuclear Weapons
310-9	(C) Index of Communications Security (COMSEC) Publications (U)
380-2	(C) SIGSEC: Defense Against SIGINT (U)
385-1	Unit Safety Management
600-8	Military Personnel Management and Administrative Procedures
600-8-1	SIDPERS Unit Level Procedures
738-750	The Army Maintenance Management System (TAMMS)

DEPARTMENT OF DEFENSE (DD) FORMS

836	Special Instructions for Motor Vehicle Drivers
1387-2	Special Handling Data/Certification

FIELD MANUALS (FM)

3-12	Operational Aspects of Radiological Defense
3-15	Nuclear Accident Contamination Control
3-21	Chemical Accident Contamination Control
3-22	Fallout Prediction
3-50	Chemical Smoke Generator Units and Smoke Operations
3-54E-TG	Trainer's Guide: 54E, NBC Specialist
3-87	Nuclear, Biological, and Chemical (NBC) Reconnaissance Decontamination Operations
5-15	Field Fortifications
5-20	Camouflage
5-25	Explosives and Demolitions
5-36	Route Reconnaissance and Classification
5-100	Engineer Combat Operations
6-20-1	Field Artillery Cannon Battalion
6-20-2	Division Artillery, Field Artillery Brigade, and Field Artillery Section (Corps)
6-50	Field Artillery Cannon Battery
6-82C1	Soldier's Manual: 82C, Field Artillery Surveyor
6-82C2/3/4-SM/TG	Trainer's Guide: Field Artillery Surveyor and Soldier's Manual (Skill Level 2/3/4)
7-7	The Mechanized Infantry Platoon and Squad
8-10	Health Service Support in a Theater of Operations
8-35	Evacuation of the Sick and Wounded
9-6	Ammunition Service in the Theater of Operations
9-13	Ammunition Handbook
9-59	Unit Operations for Support of Missile and Air Defense Gun Systems
9-207	Operation and Maintenance of Ordnance Materiel in Cold Weather (0 Degrees to Minus 65 Degrees F)
10-14	Unit Supply Operations (Manual Procedures)
10-23	Army Food Service Operations
10-24	Ration Distribution Operations
10-60	Subsistence Supply and Management in Theaters of Operations
10-63	Handling of Deceased Personnel in Theaters of Operations
10-70-1	Petroleum Reference Data
10-76Y-TG	Trainer's Guide 76Y, Unit Supply Specialist
10-76Y1/2	Soldier's Manual: 76Y, Unit Supply Specialist (Skill Level 1/2)
10-94B1/2	Food Service Specialist
10-94B3/4	Food Service Specialist
11-36K	Commander's Manual: 36K Tactical Wire Operations Specialist
11-50	Combat Communications Within the Division
11-65	High Frequency Radio Communications

57-38	Pathfinder Operations
90-2	Tactical Deception
90-3	Desert Operations
90-10	An Infantryman's Guide to Urban Combat
90-13	River Crossing Operations
100-5	Operations
101-5	Staff Officer's Field Manual: Staff Organization and Procedure
101-5-1	Operational Terms and Graphics
101-10-1	Staff Officer's Field Manual: Organizational, Technical, and Logistical Data (Unclassified Data)
101-31-1	Staff Officer's Field Manual: Nuclear Weapons Employment Doctrine and Procedures

GRAPHIC TRAINING AID (GTA)

3-6-2	NBC Warning and Reporting System

JOINT CHIEFS OF STAFF PUBLICATIONS (JCS PUB)

13, Vol 1	(S) Policy and Procedures Governing the Authentication and Safeguarding of Nuclear Control Orders (U)
13, Vol 2	(SFRD) Policy and Procedures Governing the Permissive Action Link/Coded Switch Cipher System (U)

SUPPLY BULLETINS (SB)

9-151	Safety Labeling for Storage and Shipment of Acids, Adhesives, Cleaners, Preservatives, and Other Related Materials
11-131	Vehicular Radio Sets and Authorized Installations
11-569	Replacement and Disposition of Cable Assemblies, Telephone, CX-4566/G and CX-4760/U Containing Unsuitable Contact Assemblies
38-100	Preservation, Packaging, Packing and Marking Materials, Supplies and Equipment Used by the Army
700-20	Army Adopted/Other Items Selected for Authorization/List of Reportable Items
700-21	Area Standardization of Army Equipment
740-1	Storage and Supply Activities: Covered and Open Storage
742-1	Ammunition Surveillance Procedures
746-1	Publications for Packaging Army General Supplies

SUPPLY CATALOGS (SC)

1375-95-CL-A04	Demolition Equipment Set, Explosive Initiating, Nonelectric
4910-95-CL-A72	Shop Equipment, Automotive Maintenance and Repair: Organizational Maintenance, Common No. 2, Less Power
4910-95-CL-A73	Shop Equipment, Automotive Maintenance and Repair: Organizational Maintenance, Supplemental No. 1, Less Power
4910-95-CL-A74	Shop Equipment, Automotive Maintenance and Repair: Organizational Maintenance, Common No. 1, Less Power
4933-95-CL-A07	Tool Kit, Small Arms Repairman
5180-90-CL-N01	Tool Kit, TE 50-B
5180-90-CL-N08	Tool Kit, Carpenter's Engineer Squad
5180-90-CL-N18	Refrigeration Unit
5180-90-CL-N26	Tool Kit, General Mechanic's: Automotive

5180-91-CL-R13	Tool Kit, Electronic Equipment, TK-101/G
5180-95-CL-A31	Tool Kit, Special Weapons: Organizational Maintenance, Pershing
5180-95-CL-A44	Tool Kit, Guided Missile, Mechanical Assembler; Missile Mating (Pershing PIA)
5180-95-CL-A45	Tool Kit, Guided Missile, Mechanical Assembler: Firing Site, Pershing
5975-91-CL-D01	Splicing Kit, Telephone Cable, MK-356/G
6230-97-CL-E01	Light Set, General Illumination: 25 Outlet
6545-8-CL-D27	Medical Equipment Set, Battalion Aid Station
6675-90-CL-N02	Plotting Set, Artillery Fire Control
6675-90-CL-N04	Drafting Equipment Set, Battalion: For Charts, Sketches and Overlays
7360-90-CL-N02	Range Outfit, Field, Gasoline

STANDARD FORM (SF)

153	COMSEC Material Report

NATO STANDARDIZATION AGREEMENTS (STANAG)/QUADRIPARTITE STANDARDIZATION AGREEMENTS (QSTAG)

2008/503	Bombing, Shelling, Mortaring and Location Report
2020/510	Operational Situation Reports
2047/183	Emergency Alarms of Hazard or Attack (NBC and Air Attack Only)
2070/655	Emergency War Burial Procedures
2083	Commander's Guide on Nuclear Radiation Exposure of Groups
2101/533	Principles and Procedures for Establishing Liaison
2103/187	Reporting Nuclear Detonations, Biological and Chemical Attacks, and Predicting and Warning of Associated Hazards and Hazard Areas
2104/189	Friendly Nuclear Strike Warning to Armed Forces Operating on Land
2112	Radiological Survey
2113/534	Destruction of Military Technical Equipment
2126	First Aid Kits and Emergency Medical Care Kits
2150	NATO Standards of Proficiency for NBC Defense
2154/539	Regulations for Military Motor Vehicle Movements by Road
2225	Technical Data for Handling Custodial Nuclear Weapons
2358	First Aid and Hygiene Training in NBC Operations
2865	Recording of Data for Artillery Survey Control Points
3117	Aircraft Marshalling Signals

Note. STANAGs and QSTAGs can be obtained from Naval Publications Center, 5801 Tabor Avenue, Philadelphia, PA 19120, DD Form 1425 may be used to requisition documents.

TECHNICAL BULLETINS (TB)

CML 59	Mask, Protective, Headwound, M18
MED 269	Carbon Monoxide: Symptoms, Etiology, Treatment, and Prevention of Overexposure
MED 523	Control of Hazards to Health from Microwave and Radio Frequency Radiation and Ultrasound
MED 530	Occupational and Environmental Health Food Service Sanitation
SIG 226-8	Chargers, Radiac Detector PP-1578/PD and PP-1578A/PD

SIG 226-9	Field Expedient for Charging Radiacmeters IM-93/UD and IM-147/PD
9-337	Guided Missile Systems: Corrosion Control and Treatment
9-380-101-6	Security Classification Guide for Guided Missile System, Artillery (Pershing 1A and 2)
9-1100-802-50	(CFRD) Changing Combinations on Nuclear Weapons Locking Devices (U)
9-1100-803-15	Army Nuclear Weapons Equipment Records and Reporting Procedures
9-2300-421-20	Replacement of Red Interior Domelight Blackout Lens With Blue Domelight Blackout Lens
9-4935-262-50-1	Calibration Procedure for Electrical Cable Test Set, AN/GSM-45
9-5120-202-35	Calibration Procedure for Torque Wrenches and Torque Screwdrivers (General)
9-6625-990-35	Calibration Procedure for Multimeters AN/URM-105, AN/URM-105A, AN/URM-105B, AN/URM-105C, ME-77/U, ME-77A/U, ME-77B-U, ME-77C/U, TS-297( )/U, and Triplett Model 650; and Vacuum Tube Volt-Ohmmeter Simpson Model 303
9-6625-2051-35	Calibration Procedure for Radio Test Set, AN/PRM-34
11-5820-401-20	Radio Set AN/VRC-46: Maintenance Placard
43-0121	Inspection and Certification of Radiacmeters (Dosimeters), IM-9( )/PD, IM-93( )/PD, IM-147( )/PD, and IM-185( )/PD
43-0142	Safety Inspection and Testing of Lifting Devices
43-185	Calibration and Repair Requirements for the Maintenance of Army Materiel
380-4	(C) Electronics Security Design Criteria for Noncommunication Electromagnetic Equipment (U)
385-2	Nuclear Weapons Firefighting Procedures
746-95-1	Color, Marking, and Camouflage Pattern Painting for Armament Command Equipment
750-92-1	Maintenance Expenditure Limits for Guided Missiles
750-97-28	Maintenance Expenditure Limits for Military Standard Engines (Military Design) and Outboard

TRAINING CIRCULARS (TC)

3-1	How to Conduct NBC Defense Training
6-15E1/2(JB)	Pershing Missile Crewmember Job Book, 15E (Skill Level 1/2)
6-21G1/2(JB)	Pershing Electronics Materiel Specialist Job Book, 21G (Skill Level 1/2)
6-82C1/2(JB)	Field Artillery Surveyor Job Book, 82C (Skill Level 1/2)
8-3	Field Sanitation Team Training
11-4	Handbook for AN/VRC-12 Series of Radio Sets
16-25	Ministry in a Combat Environment
19-16	Countering Terrorism on US Army Installations
21-3	Soldier's Handbook for Individual Operations and Survival in Cold Weather Areas
23-13	Crew-Served Weapon Night Vision Sight
24-1	(O) Communications-Electronics Operation Instructions, The CEOI
24-18	Communications in a "Come As You Are" War
27-1	Your Conduct in Combat Under the Law of War
32-05-2PT	Electronic Counter-Countermeasures (ECCM): Procedures for the Communicator
32-11	How to Get Out of a Jam

TECHNICAL MANUALS (TM)

3-220	Chemical, Biological, and Radiological (CBR) Decontamination
3-4230-204-12&P	Operator's and Organizational Maintenance Manual (Including RPSTL) for Decontaminating Apparatus, Portable, DS2, 1 1/2 Quart ABC-M11
3-4230-211-10	Operator's Manual: Decontaminating and Reimpregnating Kit, Individual, ABC-M113
3-6665-225-12	Operator's and Organizational Maintenance Manual: Alarm, Chemical Agent, Automatic: Portable, Manpack, M8
3-6665-260-14	Operator's, Organizational, Direct Support, and General Support Maintenance Manual: Test Set, Chemical Agent, Automatic Alarm: M74
3-6665-261-14	Operator's, Organizational, Direct Support, and General Support Maintenance Manual: Power Supply, Chemical Agent, Automatic Alarm: M10
3-6665-304-10	Operator's Manual: Area Predictor, Radiological Fallout, ABC-M52A2
3-6665-307-10	Operator's Manual: Detector Kit, Chemical Agent: M256
3-6910-227-10	Operator's Manual: Training Set, Chemical Agent Identification: Simulants, M72A2
5-200	Camouflage Materials
5-235	Special Surveys
5-241-1	Grids and Grid References
5-241-2	Universal Transverse Mercator Grid: Zone to Zone Transformation Tables
5-315	Firefighting and Rescue Procedures in Theaters of Operations
5-1080-200-10	Operator's Manual: Camouflage Screen System: Woodland Lightweight, Radar Scattering; Camouflage Screen Support System: Woodland/Desert; Camouflage Screen System: Woodland, Lightweight, Radar Transparent; Desert, Lightweight, Radar Scattering and Radar Transparent; Snow, Lightweight, Radar Scattering and Radar Transparent; and Camouflage Screen System, Snow
5-2805-203-14	Operator's, Organizational, Intermediate (Field) (Direct Support and General Support), and Depot Level Maintenance Manual: Engine, Gasoline, 6 HP (Military Standard Models DOD 4A032-1 and 4A032-2)
5-2805-257-14	Operator, Organizational, Intermediate (Field) Direct and General Support and Depot Maintenance Manual: Engine, Gasoline, 3 HP (Military Standard Models 2A016-1, 2A016-2, and 2A016-3)
5-2805-258-14	Operator, Organizational, Direct Support and General Support Maintenance Manual: Engine, Gasoline, 10 HP (Military Standard Models 2A042-2 and 2A042-3)
5-2805-259-14	Operator's, Organizational, Direct Support, and General Support Maintenance Manual: Engine, Gasoline, 20 HP (Military Standard Models 4A084-2 and 4A083-3)
5-4120-371-14	Operator's, Organizational, Direct Support and General Support Maintenance Manual for Air Conditioner, Vertical, Compact, 18,000 BTU/Hr, 208V, 3 Phase, 50/60 HZ
5-4310-227-15	Operator's, Organizational, Direct Support, General Support, and Depot Maintenance Manual: Compressor, Reciprocating, Power Driven, Air: Gasoline Engine, 15 CFM, 175 PSI
5-4310-241-15	Organizational, Direct Support, General Support, and Depot Maintenance Manual: Compressor, Reciprocating: Air, 5 CFM, 175 PSI, Hand Truck Mounted; Gasoline Engine Driven, Champion Pneumatic Model LP-512-ENG Less Engine, Champion Pneumatic Model LP-512-ENG-1 Less Engine; Receiver-Mounted, Electric Driven, Champion Pneumatic Models OEH-34-60 ENG-1, OEH-34-60-ENG-2, OEH-34-60-ENG-3, and OEH-34-60-ENG-4
5-4930-220-12	Operator and Organizational Maintenance Manual (Including RPSTL): Tank Unit, 600 Gallon Liquid Dispensing for Trailer Mounting

5-4930-228-14	Operator, Organizational, Direct Support, and General Support Maintenance Manual for Tank and Pump Unit, Liquid Dispensing for Truck Mounting
5-6115-271-14	Operator's, Organizational, Direct Support, and General Support Maintenance Manual for Generator Set, Gasoline Engine Driven, Skid Mtd, Tubular Frame, 3 KW, 3 Phase, AC, 120/280 and 120/240V, 28V DC (Less Engine) (DOD Model MEP-016A) 60 Hz, (Model MEP-021A) 400 Hz and (Model MEP-026A) DC Hz
5-6115-275-14	Operator's, Organizational, Intermediate (Field) (Direct Support and General Support) and Depot Maintenance Manual: Generator Set, Gasoline Engine Driven, Skid Mounted, Tubular Frame, 10 KW, AC. 120/208V, 3 Phase, and 120/240V Single Phase (Less Engine): DOD Models MEP-018A, 60 Hz, and MEP-023A, 400 Hz
5-6115-323-14	Operator Crew, Organizational, Intermediate (Field) (Direct Support and General Support) and Depot Maintenance Manual: Generator Set, Gasoline Engine Driven, Skid Mtd, Tubular Frame, 1.5 KW, Single Phase, AC, 120/240V, 28 V DC (Less Engine) (DOD Model MEP-015A) 60 Hz, and (Model MEP-025A) DC, 60 Hz
5-6115-332-14	Operator, Organizational, Intermediate (Field), Direct Support, General Support, and Depot Level Maintenance Manual: Generator Set, Tactical, Gasoline Engine: Air Cooled, 5 KW, AC, 120/240V, Single Phase, 120/208V, 3 Phase, Skid Mounted, Tubular Frame (Less Engine) (Model MEP-017A), Utility, 60 Hz, and (MEP-022A) Utility, 400 Hz
5-6115-365-15	Operator's, Organizational, Direct Support, General Support, and Depot Maintenance Manual (Including RPSTL): Generator Sets, Gasoline and Diesel Engine Driven, Trailer Mounted, PU-619/M
5-6115-465-12	Operator's and Organizational Maintenance Manual for Generator Set, Diesel Engine Driven, Tactical, Skid Mounted, 30 KW, 3 Phase, 4 Wire, 120/208 and 240/416V (DOD Model MEP-005A), Utility Class, 50/60 Hz, (Model MEP-104A), Precise Class, 50/60 Hz, (Model MEP-114A), Precise Class, 400 Hz Including Auxiliary Equipment, (DOD Model MEP-005AWE), Winterization Kit, Fuel Burning (Model MEP-005AWE), Winterization Kit, Electric, (Model MEP-005ALM), Load Bank Kit, and (Model MEP-005AWM), Wheel Mounting Kit
5-6115-594-14&P	Operator, Organizational, Direct Support, and General Support Maintenance Manual (Including RPSTL) for Generator Sets, Diesel Engine Driven, Trailer Mounted, PU-405A/M, PU-406B/M, PU-732/M, PU-760/M, PU-707A/M, PU-495A/G, AN/MJQ-10A, and AN/MJQ-15
5-6130-301-13&P	Operator's, Organizational, and Direct Support Maintenance Manual (Including RPSTL) for Battery Charging Distribution Panel (Model 4D-100)
5-6665-202-13	Operator's, Organizational, and Direct Support Maintenance Manual for Detecting Sets, Mine: Aural Indication, 1DV DC Operating Power: Portable Transistorized w/Case, AN/PSS-11
5-6665-293-13&P	Operator's, Organizational, and Direct Support Maintenance Manual (Including RPSTL) for Detecting Sets, Mine, Portable, Metallic and Nonmetallic, AN/PRS-7, and Detecting Set, Mine, Metallic and Nonmetallic, AN/PRS-8
5-6675-200-14	Operator's, Organizational, Direct Support, and General Support Maintenance Manual: Theodolite: Directional, 5.9 Inch Long Telescope: Detachable Tribrach w/Accessories and Tripod
5-6675-308-12	Operator's and Organizational Maintenance Manual for Position and Azimuth Determining System, AN/USQ-70
6 231	Seven Place Logarithmic Tables
9-1005-224-10	Operator's Manual for M60, 7.62-mm Machine Gun
9-1005-237-15P	Organizational, Direct Support, General Support, and Depot Maintenance Manual (Including RPSTL): Bayonet-Knife M4, M5, M5A1, M6, and M7 w/Bayonet-Knife, Scabbard, M8A
9-1005-249-10	Operator's Manual: M16A1 Rifle
9-1010-221-10	Operator's Manual: 40-mm Grenade Launcher M203

9-1115-386-12&P — Operator's and Organizational Maintenance Manual (Encl RPSTL) M266 Nuclear Warhead Section, M272 Training Warhead Section

9-1240-381-10 — Operator's Manual: Binocular M19

9-1300-206 — Ammunition and Explosive Standards

9-1375-213-12 — Operator's and Organizational Maintenance Manual (Including RPSTL): Demolition Materials

9-1425-386-L — List of Applicable Publications (LOAP) for Pershing II Field Artillery Missile System

9-2320-209-10-1 — Operation, Installation and Reference Data for 2 1/2-Ton, 6x6, M44A1 and M44A2 Series Trucks (Multifuel): Cargo: M35A1 With and Without Winch, M35A2C, M35A2, M36A2; Tank, Fuel: With and Without Winch, M49A1C, M49A2C; Tank, Water, With and Without Winch, M50, M50A2, M50A3; Van, Shop, With and Without Winch, M109A2, M109A3; Repair Shop, With and Without Winch, M185A2, M185A3; Tractor, With and Without Winch, M275A1, M275A2; Dump, With and Without Winch, M34A2; Maintenance, Pipeline Construction, M756A2, With Winch, and Maintenance, Earth Boring and Polesetting M764

9-2320-209-10-2 — Scheduled Maintenance Operator Level for 2 1/2-Ton, 6x6: M44A1 and M42A2 Series Trucks (Multifuel): Cargo: With and Without Winch, M35A1, M35A2, M35A2C, M36A2; Tank, Fuel, With and Without Winch, M49A1C, M49A2C; Tank, Water, With and Without Winch, M50A1, M50A2, M50A3; Van, Shop, With and Without Winch, M109A2, M109A3; Repair Shop, With and Without Winch, M185A2, M185A3; Tractor, With and Without Winch, M275A1, M275A2; Dump, With and Without Winch, M342A2; Maintenance, Pipeline Construction, M756A2, With Winch; and Maintenance, Earth Boring and Polesetting, M764, With Winch

9-2320-209-10-3 — Troubleshooting Operator Level for 2 1/2-Ton, 6x6: M44A1 and M44A2 Series Trucks (Multifuel): Cargo: With and Without Winch, M35A1, M35A2, M35A2C, M36A2; Tank, Fuel, With and Without Winch, M49A1C, M49A2C; Tank, Water, With and Without Winch, M50, M50A2, M50A3; Van, Shop, With and Without Winch, M109A2, M109A3; Repair Shop, With and Without Winch, M185A2, M185A3; Tractor, With and Without Winch, M275A1, M275A2; Maintenance, Pipeline Construction, M756A2, With Winch; and Maintenance, Earth Boring and Polesetting, M764, With Winch

9-2320-209-10-4 — Maintenance Operator Level for 2 1/2-Ton, 6x6, M44A1 and M44A2 Series Trucks (Multifuel); Cargo: With and Without Winch, M35A1, M35A2, M35A2C, M36A2; Tank, Fuel, With and Without Winch, M48A1C, M48A2C; Tank, Water, With and Without Winch, M50, M50A2, M50A3; Van, Shop, With and Without Winch, M109A2, M109A3; Repair Shop, With and Without Winch, M185A2, M185A3; Tractor, With and Without Winch, M275A1, M275A2; Dump, With and Without Winch, M342A2; Maintenance, Pipeline Construction, With Winch, M756A2; and Maintenance, Earth Boring and Polesetting, With Winch, M764

9-2320-209-20-1 — Scheduled Maintenance Organizational Level, 2 1/2-Ton, 6x6, M44A1 and M44A2 Series Trucks (Multifuel); Cargo, M35A1, M35A2, M35A2C, M36A2; Tank, Fuel, M49A1C, M49A2C; Tank, Water, M50A1, M50A2, M50A3; Van, Shop, M109A2, M109A3; Repair Shop, M185A2, M185A3; Tractor, M275A1, M275A2; Dump, M342A2; Maintenance, Pipeline Construction, M756A2, and Maintenance, Earth Boring and Polesetting, M764

9-2320-209-20-2-1/2 — Organizational Level 2 1/2-Ton, 6x6, M44A1 and M44A2 Series Trucks (Multifuel); Cargo, M35A1, M35A2, M35A2C, M36A2; Tank, Fuel, M49A1C, M49A2C; Van, Water, M50A1, M50A2, M50A3; Van, Shop, M109A2, M109A3; Repair Shop, M185A2, M185A3; Tractor, M275A1, M275A2; Dump, M342A2; Maintenance, Pipeline Construction, M756A2, and Maintenance, Earth Boring and Polesetting, M764

9-2320-209-20-3-1/2/3/4 — Organizational Level for 2 1/2-Ton, 6x6, M44A1 and M44A2 Series Trucks (Multifuel); Cargo, M35A1, M35A2, M36A2C, M36A2; Tank, Fuel, M49A1C, M49A2C; Van, Water, M50A1, M50A2, M50A3; Van, Shop, M109A2, M109A3; Repair Shop, M185A2, M185A3; Tractor, M275A1, M275A2; Dump, M342A2; Maintenance, Pipeline

Construction, M756A2; and Maintenance, Earth Boring and Polesetting, M764

9-2320-272-10 — Operator's Manual for Truck, 5-Ton, 6x6, M939 Series (Diesel): Chassis M939, M940, M941, M942, M943, M944, M945; Cargo, Dropside, M923 and M925; Cargo, M924 and M926; Cargo, XLWB, M927 and M928; Dump, M929 and M930; Tractor, M931 and M932; Van, Expansible, M934 and M935; and Medium Wrecker, M936

9-2320-272-10-HR — Hand Receipt Manual Covering Content of Components End Item (COEI), Basic Issue Item (BII), and Additional Authorization Lists (AAL) for Truck, 5-Ton, 6x6, M939 Series (Diesel); Chassis, M939, M940, M941, M942, M943, M944, and M945; Cargo, Dropside, M923 and M925; Cargo, M924 and M926; Cargo, XLWB, M927 and M928; Dump, M929 and M930; Tractor, M931 and M932; Van, Expansible, M934 and M935; and Medium Wrecker, M936

9-2320-272-20-1/2 — Organizational Maintenance Manual for Truck, 5-Ton, 6x6, M939 Series (Diesel); Chassis, M939, M940, M941, M942, M943, M944, M945; Cargo, Dropside, M923 and M925; Cargo, M925 and M926; Cargo, XLWB, M927 and M928; Dump, M929 and M930; Tractor, M931 and M932; Van, Expansible, M934 and M935; and Medium Wrecker, M936

9-2320-282-10 — Operator's Manual for Truck, With Crane, 10-Ton, 8x8, M1001, M1013, M1002; and Without Crane, M1014

9-2320-282-10-HR — Hand Receipt Manual Covering Contents of Components of End Item (COEI), Basic Issue Items (BII), and Additional Authorization (AAL) for Truck, Tractor, With Crane, 10-Ton, 8x8, M1001, M1013, and Without Crane, M1014; and Truck, Wrecker, With Crane, 10-Ton, 8x8, M1002

9-2320-282-20 — Organizational Maintenance Manual for Truck, Tractor, With Crane, 10-Ton, 8x8, M1001, M1013; and Without Crane, M1014; and Truck, Wrecker, With Crane, 10-Ton, 8x8, M1002

9-2330-205-14 — Operator's, Organizational, Direct Support, and General Support Maintenance Manual (Including RPSTL): Chassis, Trailer, Generator, 2 1/2-Ton, 2 Wheel, M200A1

9-2330-213-14 — Operator's, Organizational, Direct and General Support Maintenance Manual (Including RPSTL) for Chassis, Trailer, 1 1/2-Ton, 2-Wheel M103A1, M103A2, M103A3, M103A3C, M103A4, M103A4C; Trailer, Cargo, M104, M104A1, M105A1, M105A2, M105A2C; Trailer, Tank, Water, 400 Gallon, M107A1, M107A2, M107A2C, and Trailer, Van, Shop, Folding Sides, M448

9-2330-246-14&P — Operator's, Organizational, Direct Support, and General Support Maintenance Manual (Including RPSTL) for Semitrailer, Van, Electronic 6-Ton, 2-Wheel, M348A2, M348A2C, M348A2D, M348A2F, M348A2G, M348A2H, M348A2K, M348A2N, M373A2, M373A2C, M373A2D, M373A2E6, M373A2E7, M373A3, M373A4, M373A5, XM1005, and XM1007

9-2330-358-14&P — Operator's, Organizational, Direct Support, and General Support Maintenance Manual (Including RPSTL) for Semitrailer, Tactical, Dual Purpose, Breakbulk/Container Transporter, 22 1/2-Ton, M871

9-2330-364-14&P — Operator's, Organizational, Direct Support, and General Support Maintenance Manual (Including RPSTL) for Semitrailer, Van, Electronic, NBC Hardened, Tactical, 4.5-Ton, 4-Wheel, XM1006

9-4935-262-14 — Operator's, Organizational, Direct Support, and General Support Maintenance Manual: Electrical Cable Test Set, AN/GSM-45

10-412 — Armed Forces Recipe Service

10-7360-204-13&P — Operator's, Organizational, and Direct Support Maintenance Manual (Including RPSTL) for Range Outfit, Field; Gasoline M59; Burner Unit, Gasoline M2, M2A, M2 With Safety Device; and Accessory Outfit, Gasoline, Field Range With Baking Rack

10-7360-206-13 — Operator's, Organizational, and Direct Support Maintenance Manual for Kitchen, Field, Trailer Mounted, MKT-75 and MKT-75A

11-362 — Reel Units, RL-31, RL-31B, RL-31C, RL-31D, and RL-31E (Including RPSTL)

11-381	Cable Assembly CX-1065/G; Telephone Cable Assemblies CX-1606/G and CX-1512/U; Telephone Loading Coil Assembly CU-260/G; Electrical Connector Plugs U-176/G and U-226/G; and Maintenance Kit, Cable Splicing MK-640/G
11-490	Army Communications Facilities: Autodin Station and Teletypewriter Station Operating Procedures
11-490-2	Army Communications Facilities: Telecommunications Center Operating Procedures
11-2134	Manual Telephone Switchboard, SB-86/P; Installation and Operation
11-2300-351-14&P-22	Operator's, Organizational, Direct Support, and General Support Maintenance Manual (Including RPSTL) for Installation Kit, Electronic Equipment MK-1234/GRC in Truck, Utility, 1/4-Ton, 4x4, M151, M151A1, or M151A2 for Radio Sets AN/GRC-125, AN/GRC-160, AN/VRC-46, AN/VRC-53, and AN/VRC-64
11-2300-351-15-1	Installation of Radio Set, AN/GRC-106 in Truck, 1/4-Ton, 4x4, M151
11-2300-351-15-5	Installation of Radio Set AN/VRC-47 in Truck, Utility, 1/4-Ton, 4x4, M151 or M151A1
11-2300-351-15-6	Installation of Radio Set AN/VRC-49 in Truck, Utility, 1/4-Ton, 4x4, M151 or M151A1
11-2300-459-14&P-1	Operator's, Organizational, Direct Support, and General Support Maintenance Manual (Including RPSTL), Installation Kit, Electronic Equipment, MK-1813/VRC-49 and Difference Kit for Radio Set, AN/VRC-49 in Truck, Utility, 1 1/4-Ton, M882 or M892
11-2300-459-14&P-2	Operator's, Organizational, Direct Support, and General Support Maintenance Manual Including RPSTL for Installation Kit, Electronic Equipment MK-1815/GRC-106 and Difference Kit for Radio Set, AN/GRC-106 in Truck, Utility, 1 1/4-Ton, M882 or M892
11-2300-459-14&P-5	Operator's, Organizational, Direct Support, and General Support Maintenance Manual Including RPSTL for Installation Kit, Electronic Equipment MK-1817/VRC-46 and Difference Kit for Radio Set, AN/VRC-46, and Installation Kit, Remote Audio, MK-1869/GRC in Truck, Utility, 1 1/4-Ton, M882 or M892
11-4134	Manual Telephone Switchboard SB-86/P; Field Maintenance
11-5038	Control Group, AN/GRA-6
11-5543	Radiac Sets, AN/PDR-27A, AN/PDR-27C and AN/PDR-27E (Including RPSTL)
11-5805-201-12	Operator and Organizational Maintenance Manual: Telephone Set TA-312/PT
11-5805-256-13	Operator's, Organizational, and Direct Support Maintenance Manual for Telephone Set, TA-43/PT
11-5805-262-12	Operator's and Organizational Maintenance Manual: Switchboards, Telephone, Manual, SB-22/PT and SB-22A/PT and Adapter, Tone Signaling, TA-977/PT
11-5810-256-12	Speech Security Equipment TSEC/KY-57
11-5810-256-OP1	(O) Operator's Card for Speech Security Equipment TSEC/KY-57: Net Controller Operating Procedures for Communications Security Equipment
11-5810-256-OP3	Operator's Card for Speech Security Equipment TSEC/KY-57: For Use in Wheeled Vehicles
11-5810-256-OP5	Operator's Card for Speech Security Equipment TSEC/KY-57: Retransmission
11-5810-256-OP6	Operator's Card for Speech Security Equipment TSEC/KY-57; FM Secure Remote Communications
11-5810-256-OP7	Operator's Card for Speech Security Equipment TSEC/KY-57: Point-to-Point Communication
11-5810-312-12-1	TSEC/KY-57: Installation Kits (Volume 1)
11-5810-312-12-4	TSEC/KY-57: Installation Kits (Volume 4)

11-5815-334-12	Operator's and Organizational Maintenance Manual for Radio Teletypewriter Sets, AN/GRC-142, AN/GRC-142A, AN/GRC-142B, AN/GRC-142C, AN/GRC-142D, AN/GRC-142E, AN/GRC-122, AN/GRC-122A, AN/GRC-122B, AN/GRC-122C, AN/GRC-122D, and AN/GRC-122E
11-5820-401-10-1	Operator's Manual: Radio Sets AN/VRC-42, AN/VRC-43, AN/VRC-44, AN/VRC-45, AN/VRC-46, AN/VRC-47, AN/VRC-48, and AN/VRC-49 (Used Without an Intercom System)
11-5820-401-10-1-HR	Hand Receipt Manual Covering Radio Sets
11-5820-401-10-LD-5	Plastic Laminated Condensed Operating Instructions for Radio Set AN/VRC-46
11-5820-401-10-LD-6	Plastic Laminated Condensed Operating Instructions for Radio Set AN/VRC-47
11-5820-401-10-2	Operator's Manual: Radio Sets AN/VRC-12, AN/VRC-43, AN/VRC-44, AN/VRC-45, AN/VRC-46, AN/VRC-47, AN/VRC-48, and AN/VRC-49 (Used Without an Intercom System)
11-5820-401-10-2-HR	Hand Receipt Manual Covering End Item/Components of End Items (COEI), Basic Issue Items (BII), and Additional Authorization List (AAL) for Radio Sets AN/VRC-12, AN/VRC-43, AN/VRC-44, AN/VRC-45, AN/VRC-46, AN/VRC-47, AN/VRC-48, and AN/VRC-49 (Used Without Intercom Systems)
11-5820-401-12	Radio Sets (Including RPSTL)
11-5820-610-14	Operator's, Organizational, Direct Support, and General Support Maintenance Manual: Radio Terminal Set, AN/TRC-133A
11-5985-262-14	Operator's, Organizational, Direct Support, General Support, and Depot Maintenance Manual for Antenna AS-1729/VRC
11-5985-357-13	Operator's, Organizational, and Direct Support Maintenance Manual for Antenna Group OE-254/GRC
11-6130-266-15	Operator's, Organizational, Direct Support, General Support, and Depot Maintenance Manual (Including RPSTL) for Power Supply PP-6224/U and PP-6224A/U
11-6625-203-12	Operator's and Organizational Maintenance Manual: Multimeter, AN/URM-105 and AN/URM-105C
11-6665-214-10	Operator's Manual: Radiacmeters IM-93/PD, IM-93/UD, IM-93A/UD, and IM-147/PD
38-L22-15-2	Functional User's Manual for Division Logistics System (DLOGS): Class IX (Repair Parts) Supply System Supply Operating Procedures: Using Unit Procedures
39-0-1A	Numerical Index to Joint Nuclear Weapons Publications (Including Related Publications) (Army Supplement)
743-200-1	Storage and Materials Handling
743-200-2	Storage Modernization
743-200-3	Storage and Materials Handling
750-244-2	Procedures for Destruction of Electronics Material to Prevent Enemy Use
750-244-6	Procedures for Destruction of Tank-Automotive Equipment to Prevent Enemy Use

TABLES OF ORGANIZATION AND EQUIPMENT (TOE)

06600J300	FA Brigade, Pershing
06626J300	HH&SB, FA Battalion, Pershing
07015H010	Infantry Battalion, E/W 106
63075J300	Support Maintenance Battalion, Pershing Brigade

MISCELLANEOUS

Manual for Courts Martial (MCM) (UCMJ)

PROJECTED PUBLICATIONS

Projected publications are sources of additional information that are scheduled for printing but not yet available.

ARMY TRAINING AND EVALUATION PROGRAM (ARTEP)

6-625	Pershing II Field Artillery Battalion

FIELD MANUALS (FM)

3-3	NBC Contamination Avoidance
3-4	NBC Protection
3-5	NBC Decontamination
3-100	NBC Operations
100-16	Support Operations, Echelons Above Corps

TECHNICAL BULLETIN (TB)

9-1425-386-14-1	Color and Marking of Pershing II Missile and Support Equipment

TECHNICAL MANUALS (TM)

3-4240-298-20&P	Collective Protection Equipment, Pershing II, Consisting of Entrance, Protective, Pressurized, Collapsible, XM14; Filter Unit, Gas Particulate, 200 CFM, 208V, 400 Hz (Encl RPSTL)
3-4240-299-23&P	Static Frequency Converter: XM5 Modular Collective Protection Equipment (Encl RPSTL)
9-1115-230-14&P	(CRD) Operator's Through General Support Maintenance Manual (Including RPSTL): Trainer M271 (U)
9-1410-387-12	Operator's and Organizational Maintenance Manual: Pershing II Surface Attack Guided Missile
9-1425-386-10-HR	Pershing II Hand Receipt and Inventory List
9-1425-386-10-1	Pershing II Weapon System (System Description)
9-1425-386-10-2/1	Pershing II Weapon System (Firing Position Procedures) Volume 1 Tabulated Procedures
9-1425-386-10-2/2	Pershing II Weapon System (Firing Position Procedures) Volume II Detailed Procedures for EL OP and PCC CM1
9-1425-386-10-2/3	Pershing II Weapon System (Firing Position Procedures) Volume III Detailed Procedures for CM 1 Through CM4 and PCC CM2
9-1425-386-10-3/1	Pershing II Weapon System (Missile Assembly)
9-1425-386-10-3/2	Pershing II Weapon System (Missile Disassembly)
9-1425-391-14	Operator, Organizational, and DS/GS Maintenance Manual: Shelter, Electrical Equipment, Facilitized
9-1430-388-12	Operator and Organizational Maintenance Manual: Reference Scene Generation Facility
9-1430-392-10	Operator's Manual: Platoon Control Central, Guided Missile, Transportable, and Battery Operations Center, Guided Missile, Transportable
9-1430-392-20	Organizational Maintenance Manual: Platoon Guided Missile, Transportable, and Battery Operations Center, Guided Missile, Transportable
9-1430-393-14	Operator, Organizational, and DS/GS Maintenance Manual: Power Distribution System, Guided Missile System
9-1440-389-10	Operator's Manual: Erector Launcher, Guided Missile, Semitrailer Mounted XM 1003
9-1440-389-20	Organizational Maintenance Manual: Erector Launcher, Guided Missile, Semitrailer Mounted XM 1003

9-1450-387-13	Operator's Through Direct Support Maintenance Manual: Rear Area Power Unit
9-1450-396-14	Operator's Through Direct Support Maintenance Manual: Ground Handling Equipment
9-2320-279-10	Operator's Manual: Truck, Tractor, 10-Ton, 8x8, M983 HEMTT
9-2320-279-10-HR	Hand Receipt Manual: 10-Ton HEMTT
9-2320-279-20	Organizational Maintenance Manual: 10-Ton HEMTT
9-4935-387-14	Operator's Through General Support Maintenance Manual: Test Set, Safe and Arm Recoding
9-4935-394-14	Operator's Through General Support Maintenance Manual: Mechanical and Electrical Shop Sets
9-4935-395-14	Operator's Through General Support Maintenance Manual: Preservation and Packaging Shop, Supply Office, and Repair Parts Shop
9-4935-396-14	Operator's Through General Support Maintenance Manual: Electronic Filter Test Set
9-6920-387-14	Operator's Through General Support Maintenance Manual: First Stage Trainer MT-1, Second Stage Trailer MT-2, G&CA Section Trainer MT-3, Radar Section Trainer MT-6, and Missile Simulator ES-1
9-6920-388-14	Operator's Through General Support Maintenance Manual: RSGF Trainer
9-6920-389-14	Operator's Through General Support Maintenance Manual: Simulator, Missile/Erector Launcher/Ground Integrated Electronic Unit
9-6920-392-14	Operator's Through General Support Maintenance Manual: Platoon Control Central Guided Missile System Trainer
9-8140-395-14	Operator's Through General Support Maintenance Manual: Shipping and Storage Containers

☆U.S. GOVERNMENT PRINTING OFFICE: 1985 537 031 20062

By Order of the Secretary of the Army:

JOHN A. WICKHAM, JR.
General, United States Army
Chief of Staff

Official:

DONALD J. DELANDRO
Brigadier General, United States Army
The Adjutant General

DISTRIBUTION:

Active Army and USAR: To be distributed in accordance with Da Form 12-11A, Requirements for Field Artillery Tactics (Qty rqr block no. 39); Field Artillery Battalion, Pershing (Qty rqr block no. 43); and DA Form 12-32, Requirements for Pershing (Qty rqr block no. 607).

ARNG: None.

Additional copies may be requisitioned from the US Army Adjutant General Publications Center, 2800 Eastern Boulevard, Baltimore, MD 21220.

Made in the USA
San Bernardino, CA
13 December 2015